Hall's assessment of how the UNHCR, IOM and UNDP have adapted their mandates and practices to address the climate change agenda is a welcome breakthrough in what we know about how international organizations change and why such change varies across different types of organizations. This will be important reading for any student of international organizations and a major contribution to policymakers who want to know how we can reform these major organizations to address one of the most pressing global issues of our time.

Catherine Weaver, Associate Professor and Distinguished Scholar, The University of Texas at Austin, USA

Hall provides a perceptive critique of why and how mandates evolve within international organisations. Her book is a must-read for all seeking to ensure that our global institutions remain fit for purpose.

Sam Daws, Director, Project on UN Governance and Reform, Oxford University, UK

This book provides fascinating insights into how international development, migration and humanitarian organizations are responding to the challenge of climate change

Professor Jane McAdam, Scientia Professor of Law and Director of the Andrew & Renata Kaldor Centre for International Refugee Law, University of NSW, Australia

Displacement, Development, and Climate Change

Although scholars are examining the complexity and fragmentation of the climate change regime, they have not examined how our existing international development, migration, and humanitarian organizations are dealing with climate change. This book addresses that challenge. An increase in extreme weather events, global temperatures, and higher sea levels may lead to displacement and migration, and affect many dimensions of the economy, and society.

Focusing on three institutions: the United Nations High Commissioner for Refugees, the International Organization for Migration, and the United Nations Development Programme, none of which was established with a mandate for climate change, the book asks: How have these intergovernmental organizations responded to climate change? And are they moving beyond their original mandates? It traces their responses to climate change in their rhetoric, policy, structure, operations, and overall mandate change. Hall argues that international bureaucrats can play an important role in mandate expansion, often deciding whether and how to expand into a new issue-area and then lobbying states to endorse this expansion. They make changes in rhetoric, policy, structure, and operations on the ground, and therefore forge, frame, and internalize new issue-linkages.

This book helps us to understand how institutions established in the twentieth century are adapting to a twenty-first-century world. It will be of great interest to scholars and students of international relations, development studies, environmental politics, international organizations, and global governance, as well as international officials.

Nina Hall is a postdoctoral fellow at the Hertie School of Governance. Her research explores how international organizations are evolving in the twenty-first century.

Global Institutions

Edited by Thomas G. Weiss
The CUNY Graduate Center, New York, USA
and Rorden Wilkinson
University of Sussex, Brighton, UK

About the series

The "Global Institutions Series" provides cutting-edge books about many aspects of what we know as "global governance." It emerges from our shared frustrations with the state of available knowledge—electronic and print-wise, for research and teaching—in the area. The series is designed as a resource for those interested in exploring issues of international organization and global governance. And since the first volumes appeared in 2005, we have taken significant strides toward filling conceptual gaps.

The series consists of three related "streams" distinguished by their blue, red, and green covers. The blue volumes, comprising the majority of the books in the series, provide user-friendly and short (usually no more than 50,000 words) but authoritative guides to major global and regional organizations, as well as key issues in the global governance of security, the environment, human rights, poverty, and humanitarian action among others. The books with red covers are designed to present original research and serve as extended and more specialized treatments of issues pertinent for advancing understanding about global governance. And the volumes with green covers—the most recent departure in the series—are comprehensive and accessible accounts of the major theoretical approaches to global governance and international organization.

The books in each of the streams are written by experts in the field, ranging from the most senior and respected authors to first-rate scholars at the beginning of their careers. In combination, the three components of the series—blue, red, and green—serve as key resources for faculty, students, and practitioners alike. The works in the blue and green streams have value as core and complementary readings in courses on, among other things, international organization, global governance, international law, international relations, and international political economy; the red volumes allow further reflection and investigation in these and related areas.

The books in the series also provide a segue to the foundation volume that offers the most comprehensive textbook treatment available dealing with all the major issues, approaches, institutions, and actors in contemporary global governance—our edited work *International Organization and Global Governance* (2014)—a volume to which many of the authors in the series have contributed essays.

Understanding global governance—past, present, and future—is far from a finished journey. The books in this series nonetheless represent significant steps toward a better way of conceiving contemporary problems and issues as well as, hopefully, doing something to improve world order. We value the feedback from our readers and their role in helping shape the on-going development of the series.

A complete list of titles appears at the end of this book. The most recent titles in the series are:

UN Security Council Reform (2016)
by Peter Nadin

International Organizations and Military Affairs (2016)
by Hylke Dijkstra

The International Committee of the Red Cross (2nd edition, 2016)
by David P. Forsythe and Barbara Ann J. Rieffer-Flanagan

The Arctic Council (2016)
by Douglas C. Nord

Human Development and Global Institutions (2016)
by Richard Ponzio and Arunabha Ghosh

NGOs and Global Trade (2016)
by Erin Hannah

Brazil as a Rising Power (2016)
edited by Kai Michael Kenkel and Philip Cunliffe

Displacement, Development, and Climate Change

International organizations moving beyond their mandates

Nina Hall

Routledge
Taylor & Francis Group
LONDON AND NEW YORK

First published 2016
by Routledge
2 Park Square, Milton Park, Abingdon, Oxon OX14 4RN

and by Routledge
711 Third Avenue, New York, NY 10017

Routledge is an imprint of the Taylor & Francis Group, an informa business

British Library Cataloguing in Publication Data
A catalogue record for this book is available from the British Library

Library of Congress Cataloging in Publication Data
Names: Hall, Nina, 1985-
Title: Displacement, development, and climate change : international organizations moving beyond their mandates / Nina Hall.
Description: Abingdon, Oxon ; New York, NY : Routledge, 2016.
Identifiers: LCCN 2015042485| ISBN 9781138190535 (hbk) | ISBN 9781315639758 (ebk) | ISBN 9781138190542 (pbk)
Subjects: LCSH: Climate change mitigation–International cooperation. | Climatic changes–International cooperation.
Classification: LCC QC903 .H2154 2016 | DDC 363.738/747–dc23
LC record available at http://lccn.loc.gov/2015042485

ISBN: 978-1-138-19053-5 (hbk)
ISBN: 978-1-138-19054-2 (pbk)
ISBN: 978-1-31563-975-8 (ebk)

Typeset in Times New Roman
by Taylor & Francis Books

To Trish and Philip

Contents

List of figures

Acknowledgments

In 2009 as the world was looking towards the United Nations Framework Convention on Climate Change in Copenhagen, I began this research. As a New Zealander I was concerned by the possible impacts of climate change on people living in the Pacific Islands, and attended a conference on the issue in May 2009 in Wellington, at the Institute of Policy Studies, Victoria University. This conference and initial conversations with Jane McAdam triggered my interest in how existing international migration, refugee, and development institutions were responding to climate change. I explored these questions through a doctorate at the University of Oxford, where I met and learned from many and continued this research as postdoctoral fellow at the Hertie School of Governance.

Researching this book took me from Kenya to Copenhagen, Sydney to Oslo. This research was possible thanks to the support of a Rhodes Scholarship and grants from St Antony's College, the Department of Politics and International Relations, and the Gilbert Murray Trust. I owe a great deal to all those who contributed to my work as interview participants, from Kenyan government officials to NGO staff. A special thank you to the staff of UNHCR, UNDP, and IOM for your time, openness, and reflections on these evolving policy debates. Understanding the work that you do and how your institutions are adapting to deal with twenty-first-century problems drove this research forward.

I am immensely grateful to Alexander Betts and Duncan Snidal, my doctoral supervisors, for their advice and direction. Conversations with Alex challenged my assumptions (in particular that climate change would *cause* migration), stretched my thinking, and guided my work. Duncan refined my theoretical contributions, particularly through doctoral workshops, which were a critical part of my intellectual development. Ngaire Woods and Catherine Weaver gave constructive and excellent feedback as examiners of my doctoral thesis, which provided the starting point for this book. Thanks also to Jane McAdam, Gil Loescher, Guy Goodwin-Gill, Andrew Hurrell, and Jennifer Welsh, who also gave insightful advice.

My work benefitted greatly from the feedback and support of other peers and colleagues. They include: Dani Marinova, Mark Dawson, Markus Jachtenfuchs, Tom Hale, Dave Blagden, Francesca Giovannini, Emily Jones, Amy King, Michael Manulak, and Diarmuid Torney. The editors of the Global Institutions series, Thomas Weiss and Rordern Wilkinson, and their colleagues at Routledge were thoughtful and efficient at getting the book through the production process. I also thank the anonymous peer reviewers who provided constructive feedback on this work in progress. Lisa Schmid, my research assistant, was a superb copy editor of the final manuscript.

Friends and family have supported me in many ways through this process. Thanks to a wide circle of friends who offered a bed, a meal, an outdoor expedition, or your curious ear and enthusiastic encouragement throughout the process. There are many of you. Thanks in particular to: Garima Mohan, Olivia Gippner, Lea Neubert, Jess Pettersen, Malebogo Ngoepe, Katrin Scholz, Christopher Arpi, Spencer Stoner, Liz Fouksman, Alysia Garmulewicz, Beth Foley, Rosalie Hughes, Kitty Garden, Jess Panegyres, Aleisha Keating, Priscilla Solano, Dan Tremewan, Lucy Cassels, Jess Auerbach, Kate Vyborny, Emma Preston, Julia Shaw, Lucy Hawcroft, Laura Cleary, Lillian Fougère, Steffen Lohrey, Anne Buffardi, Francis Vergunst, Caitlin Prentice, Sarah Silver, Florian Strob and the Shannon Commonwealth. A large thank you to my father, Philip Tremewan, who was a meticulous proofreader: your comments strengthened this work throughout its multiple iterations. Finally, Philip, Trish (my mother), and Ralph (my brother), all provided incredible support throughout this journey.

Abbreviations

ADB	Asian Development Bank
ASEAN	Association of Southeast Asian Nations
AWG-LCA	Ad Hoc Working Group on Long-term Cooperative Action under the Convention
CARE	Cooperative for Assistance and Relief Everywhere
CCEMA	Climate Change, Environment and Migration Alliance
COP	Conference of the Parties
CRMA	Climate Risk Management and Adaptation Strategy
DFID	United Kingdom's Department for International Development
DRR	Disaster Risk Reduction
EACH-FOR	Environmental Change & Forced Migration Scenarios
ECOSOC	Economic and Social Council
ECOWAS	Economic Community of West African States
EPTA	Expanded Programme for Technical Assistance
EU	European Union
ExCom	Executive Committee of the Programme of the UNHCR
FAO	Food and Agriculture Organization
G77	Group of 77
GEF	Global Environment Facility
IASC	Inter-Agency Standing Committee
ICEM	Intergovernmental Committee for European Migration
IDPs	Internally Displaced Persons
IFRC	International Federation of Red Cross and Red Crescent Societies
IIED	International Institute for Environment and Development
ILO	International Labour Organization
IMF	International Monetary Fund

IOM	International Organization for Migration
IPCC	International Panel on Climate Change
LDCs	Least Developed Countries
NAPA	National Adaptation Programme of Action
NATO	North Atlantic Treaty Organization
NGO	Non-governmental organization
NRC	Norwegian Refugee Council
OAU	Organization for African Unity
OCHA	Office for the Coordination of Humanitarian Affairs
OECD	Organisation for Economic Co-operation and Development
OPEC	Organization of Petroleum Exporting Countries
PICMME	Provisional Intergovernmental Committee for the Movement of Migrants from Europe
UN	United Nations
UN Women	United Nations Entity for Gender Equality and the Empowerment of Women
UNCBD	UN Convention on Biological Diversity
UNCED	Rio United Nations Conference on Environment and Development
UNDP	United Nations Development Programme
UNEP	United Nations Environment Programme
UNFCCC	United Nations Framework Convention on Climate Change
UNFPA	United Nations Population Fund
UNHCR	United Nations High Commissioner for Refugees
UNICEF	United Nations Children's Fund
UNIDO	United Nations Industrial Development Organization
UN-REDD	United Nations Reductions of Emissions due to Deforestation and forest Degradation
UNU	United Nations University
UNU-EHS	United Nations University, Institute of Environmental and Human Security
WFP	World Food Programme
WHO	World Health Organization
WTO	World Trade Organization

Introduction

- **Theorizing mandate change in international organizations**
- **Studying mandate change**
- **Research design**
- **Structure of the book**
- **Conclusion**

In 1951 in Geneva, at the heart of Europe recovering from a devastating war, states negotiated the Convention Relating to the Status of Refugees, a landmark agreement which offered protection to those forced to flee across borders due to persecution. States tasked the newly created United Nations High Commissioner for Refugees (UNHCR) with overseeing and supervising this convention. A small office was established in Geneva for international refugee lawyers to advise states and ensure refugee law was adhered to. In that same year, states established another new international organization: the Provisional Intergovernmental Committee for the Movement of Migrants from Europe (PICMME). Its purpose was to organize the relocation of thousands of labor migrants from post-war Europe to the Americas, Australia, and New Zealand. To perform these tasks PICMME had a fleet of ships and officers with experience in the mass movement of migrants across the Atlantic and Pacific Oceans. These two organizations were just the tip of the iceberg of a growing population of intergovernmental organizations, created in the aftermath of World War II.

Fast forward more than sixty years and these two organizations, along with many others, still exist. Yet the world has changed dramatically and the problems they were initially established to address are no longer the same. A globalized and interconnected world has brought new challenges ranging from the global financial crisis to climate change. Many intergovernmental organizations were established on the premise that global challenges could be siloed into discrete issue-areas such as migration, health, and trade. But today the issues are seen as so interdependent that we can no longer address one issue without also targeting many others. Humanitarian crises, for instance, are intertwined with issues such as food security, development, climate change, and gender equality. Given this changing international context, how are intergovernmental institutions established in the post–World War II era adapting?

This book focuses on one critical new challenge: climate change. Climate change is predicted to lead to an increased intensity and frequency of natural disasters from droughts to floods and higher sea-levels. Its impacts are already being felt and seen across the globe: from extreme weather events, such as the cyclone which destroyed much of Vanuatu in 2014, to intense and prolonged droughts in the horn of Africa. In addition, an increase in extreme weather events and higher sea levels may lead to displacement and migration.[1] As global temperatures rise this affects many dimensions of the economy and society, from agriculture to public health. Climate change is also a challenge to the existing development paradigm—development organizations should be promoting sustainable, green growth rather than following the fossil fuel intensive industrialization model of the west.[2] Development actors must also enable countries to adapt and prepare for the likely increase in slow-onset and fast-onset disasters.[3]

Climate change also has dramatic effects on developed countries as can be seen from the powerful hurricanes hitting New Orleans and the subsequent flooding, the heat waves that now regularly feature in European summers, and the destructive bush-fire seasons in Australia. But these countries have the resources to cope, whereas developing countries most often do not.

In the past decade many international organizations have highlighted the disastrous effects of climate change on developing countries. Many organizations, from Greenpeace to Oxfam, highlighted how climate change may lead to climate change "refugees" (a popular but legally incorrect term, as will be discussed in this book) as people are forced to flee their homes. The media also picked up on the stories of low-lying islands, such as Kiribati and Tuvalu in the Pacific, which would be inundated by a sea-level rise of several meters. People living in delta areas, particularly in Bangladesh, are also at risk of flooding. Many feared that climate change would drive millions from their homes—displaced by flooding, storms, droughts, or sea-level rise.[4] Concerned civil society organizations, academics, and non-governmental organizations (NGOs) argued for new international protection frameworks to cover those forced to move due to climate change.[5] Although a large literature has emerged examining the links between climate change, migration, and displacement (and will be discussed in Chapter 2) there has been little examination of how existing development, migration, and refugee organizations have adapted to the new reality of climate change. Are they able to assist those worst affected by climate change—be they "climate change migrants," climate refugees, or those left behind—within their existing operations and activities?

In fact, we know relatively little about how international development, migration, and humanitarian organizations are responding to climate change. Global environmental governance scholars have focused on global environmental organizations and how they are responding to climate change. They have also explored the increasing complexity and fragmentation of the climate regime as it broadens to include a range of new issues and new actors from NGOs to interest groups and intergovernmental organizations.[6] They have identified why international organizations may "band-wagon" on the climate regime.[7] They have pointed out the benefits institutions can gain, in particular new material and ideational resources, and the role that secretariats play in managing regime interplays.[8] However scholars have not sufficiently examined how existing *non-environmental* institutions are adapting to climate change and what impact this is having on their mandates. Are they moving beyond their mandates?

This book focuses on three institutions: the UNHCR, the International Organization for Migration (IOM), and the United Nations Development Programme (UNDP). It asks: How have these intergovernmental organizations responded to climate change? And how can we explain variation in their responses? These three institutions have a leading role to play in assisting developing countries, who are predicted to be the most vulnerable to climate change. This is particularly important as there is no new global "adaptation" organization tasked with helping developing countries prepare for and deal with the impacts of climate change. In fact, the United Nations (UN) Secretary General has argued adaptation should be included in the work of all UN agencies and programs and it is thus imperative to understand if and how they have done so.

In addition, it is important to understand how these three existing organizations are adapting because states rarely create new institutions, or disestablish old ones.[9] Given that organizational change can vary from the ad-hoc and superficial to the sustainable or transformative, we need to ask what sorts of changes are occurring and why. Demands for transformations in global governance often overlook the value of a degree of inertia. If our international organizations were to constantly change their mandates and tasks they would be unable to deliver effectively in areas where they hold expertise. Change may mean neglecting, or even abandoning, tasks they were originally designed to do. International organizations have to walk a tightrope between stasis and change. There are thus wider implications of this research: When and what sort of institutional change should we advocate for?

This book's central argument is that international bureaucrats can play a decisive role in mandate expansion. They often decide whether and how to expand into a new issue-area and then lobby states to endorse this expansion. In particular, staff of international organizations make changes in rhetoric, policy, structure, and operations on the ground. Through these actions they forge, frame, and internalize new issue-linkages and use these to advocate for expansion. The core contribution of this book is untangling how international organizations see and adapt their mandate to a changing external environment, when states do not give them clear direction.

Theorizing mandate change in international organizations

This book is in conversation with a long-standing international relations debate on the autonomy and influence of international organizations. Much of the literature takes a statist view—that states determine whether to "Use, Select, Change or Create" institutions to deal with new issues.[10] In contrast, other recent scholarship has emphasized the role of international bureaucrats in creating institutions and shaping regime overlaps.[11] This book contributes to these contemporary debates by illustrating how international organizations can lobby states to expand their mandates and even bypass states' approval. Mandates are set by a collective of states, who govern and finance these institutions. Yet states do not typically delegate every task or action, but rather expect an organization to use its expertise and respond to new issues autonomously. In addition, international organizations have some wiggle room: the exact parameters of a mandate are often ambiguous.[12] International organizations can exploit differences between states over what an organization should or should not do. Thus an organization may extend the limits of its mandate by engaging with a new issue, and lobbying states to support this expansion officially.

In doing so, this book builds on and contrasts with the work of principal agent scholars. Principal agent theory starts from the assumption that states (principals) create intergovernmental organizations (agents) to perform particular tasks on their behalf. These agents offer specialized expertise and services which would be costly for any one principal to provide.[13] International organizations may also enable cooperation by: facilitating collective decision making, resolving disputes, and enhancing credibility of decisions by locking in particular policies.[14] Although states delegate international organizations limited autonomy to perform these tasks, these institutions can only go so far beyond, or against, state interests as they will lose financial and political support. According to this account organizational expansion occurs either if states delegate a new task to an international organization (delegation) or if an organization strays off its delegated mandate into a new area, so-called "agency slack."

The principal agent literature is concerned by agency slack, and sees it as a problem to be constrained. Scholars have identified how states monitor international organizations and ensure they do not stray too far from their mandate. They have shown how the degree of discretion delegated, proximity of member state preferences, heterogeneous principal preferences, voting structures, asymmetry of information, and the costs of monitoring, explain why one organization will expand and another does not.[15] Principals can also use a number of control mechanisms to ensure organizations do not engage in agency slack. They may simply delegate little discretion to the agent.[16] They may also monitor and enforce reporting requirements through "police patrols" or "fire alarm" oversights.[17] States may establish institutional checks or sanction an institution by controlling the budget, staff appointments, and overriding agency behavior through new legislation. These control mechanisms enable principals to ensure compliance with their preferences, but may be costly to implement. States face a trade-off between the costs of monitoring and perceived agency losses. A principal may find monitoring an agent too costly and/or be ineffective at controlling an agent.[18] Thus an agent will gain greater slack if its principals decide not to—or cannot—effectively control it.

These are all useful insights; however they do not examine how mandate evolution may occur over time as a result of persistent lobbying by bureaucrats of international organizations. I acknowledge that states set important parameters of action for international organizations—after all only they can stipulate the founding mandate of an international organization. However, international organizations are not completely restricted by their member states. They may choose to expand into new issue-areas even if states are not supportive and actively voice their dissent. Thus we may see an organization gaining greater scope, and taking on a range of new issues, without any collective agreement from states.

In addition, this book highlights two factors which shape whether and how an international organization responds to a new issue-area: issue-linkages and staff perception of their mandate. Essentially, it proposes that organizational expansion is most likely to occur when staff perceive a new issue to link strongly to their organizational mandate. It builds on international relations scholarship emphasizing the role of international bureaucracies, staff, and leaders in organizational change.[19] Let's look at these two claims in turn and how they build on and contribute to existing scholarship.

Issue-linkages

An organization's decision to engage with a new issue depends firstly on the nature of that issue, and how relevant it is to its mandate. Twenty-five years ago it would have seemed odd for an international humanitarian organization to engage in debates over how to reduce greenhouse gas emissions in the atmosphere. The catch is that issues evolve in complexity as we become increasingly aware of how they impact on other issues—and climate change is now seen to impact a huge number of areas, including issues of direct consequence to humanitarian and development organizations. It is often classified as a "wicked" problem, an issue which is inherently difficult to resolve because of its complexity and interdependency with other issues.

The question of if and when one issue "spills over" and becomes relevant to other regimes is a central concern for scholars of regimes and international organizations. There are many examples of, and a growing literature on, emerging issue-linkages that bridge distinct regimes such as: development and migration,[20] gender and security,[21] or trade and environment.[22] One shortcoming of the existing literature is that it does not examine the strength of these issue-linkages.[23] I propose that the more widely accepted, or perceived, an issue-linkage is the stronger it is.

Importantly, I distinguish between substantive and tactical issue-linkages. A substantive issue-linkage is a "real or perceived-to-be-real causal relationship between two discrete issue areas."[24] For instance, we can say there is a substantive issue-linkage between climate change and health because, for example, climate change shifts areas where malarial mosquitos can live and breed, hence causally impacting on public health. Substantive issue-linkages differ from tactical issue-linkages, which occur when issues are combined in interstate negotiations through conditionality to facilitate bargaining. An example of a tactical issue-linkage is that developing states may make their pledges to reduce greenhouse gas emissions dependent on whether donor states provide climate financing.

Substantive issue-linkages do not exist on their own accord—they must be established and actively promoted by actors who see a reason for linking two discrete issue areas. They are intersubjectively constituted and maintained by a range of international actors including: international organizations, international NGOs, and states. Intersubjectivity means that no single actor can on their own create a strong issue-linkage. Rather, the creation of strong substantive issue-linkages requires a consensus or agreement between a number of influential actors that there is a causal link between two discrete areas. A strong issue-linkage will be highly visible and widely supported by international actors in at least one of the two regimes being linked. The difference between a weak issue-linkage and a strong issue-linkage is often the presence of a successful epistemic community that endorses and promotes the issue-linkage.[25]

Strong issue-linkages cannot be reduced to a single indicator or criterion. It is difficult to measure the degree of institutional support and/or consensus behind a substantive issue-linkage. I use the following indicators in this book to identify when an issue-linkage becomes strong: a significant number of international NGOs and international organizations elaborating policy or advocacy material about an issue-linkage; high citations and circulation of literature substantiating the issue-linkage; presence of interagency working groups dedicated to the issue-linkage; and senior politicians who endorse the link. These indicators will be used in chapter two to identify when climate change becomes linked to development and migration respectively.

This conception of evolving issue-linkages stands in contrast to functionalist theories of "spillovers." Functionalists have highlighted how international organizations should be created to pursue specific technical tasks—such as coordinating trade or communications.[26] Neo-functionalists, building on this theory and the empirical success of the European Union (EU) common market, propose that interdependent economic interests create a demand for institutions to deliver cooperative outcomes in other areas.[27] Once regional economic institutions are created there is an inherent tendency for cooperation to "spill over" into other issue areas. For functionalists, cooperation on one issue between states will trigger further cooperation on other issues and leads to organizational expansion. This occurs as the nature of a problem evolves or the original problem that an international organization was created for is solved. However, functionalists consider international organizations as passive in this process: they are independent variables and organizational change is determined by exogenous factors.[28] I argue, and elaborate next, that international organizations are autonomous and play an active role in instigating and elaborating issue-linkages.

Staff perception of issue-linkage

The mere presence of an issue-linkage between a new issue and an organization's mandate does not entail that an organization will expand and engage with it. Staff perception of their mandate will shape if and how they respond. In making this argument, the book builds on sociological institutionalism and principal agent accounts which are often presented as competing approaches. Yet theorists in both traditions claim that bureaucrats in international organizations develop their own identities and preferences and, due to this, influence organizational behavior. I suggest that international bureaucrats will decide which issues are closely linked to their mandate and choose if and how to engage with them.

Principal agent scholars and sociological institutionalists agree that international organizations, staffed by international, professional bureaucrats, will develop preferences independently of member states.[29] These staff share a common professional identity, and will be loyal to their international organization as they depend on it for their job security and promotion. Independent professional bureaucrats will be dedicated to the "knowledge of problem-based aspects of the international organization's mission" and will create "epistemic communities."[30] Scholars differ on what factors shape staff decision to expand. Some highlight the role of culture, identity, and epistemic beliefs (namely sociological institutionalists) whereas others (many principal agent scholars) tend to focus on availability of resources or perception of states' reaction.

I suggest that staff will favor expansion if a new issue resonates with the core role and identity of an organization. International organizations will tend to expand in both size and scope as staff "try to square their rationalized abstractions of reality with facts on the ground."[31] This is because "conscientious bureaucrats very quickly recognize that to accomplish a great many ambitious social tasks they need to reach outside the narrow compartments in which we place them."[32] For expansion to occur staff need to perceive a strong issue-linkage between their core issue-area and the new issue-area. "[They] will be more receptive to new goals and more active in implementing them when they believe that these goals flow logically from the organization's own expert knowledge" than if they come from outside or are perceived to be tangential.[33]

Staff may not always favor expansion. This is because institutional action is based on "identifying the normatively appropriate behavior rather than calculating the return expected from alternative choices."[34] This is particularly true when organizations face novel and politically important scenarios.[35] Individuals in organizations will decide on a course of action based on their interpretation of existing norms and rules, rather than rationally calculating the cost-benefit return expected from their choices. Thus expansion will occur if international bureaucrats perceive that they should play a role in the new issue-area.[36] If a new issue arises organizations will ask: Does it fit with the existing tasks, identity, or culture of the organization?

This explanation contrasts with other accounts of staff preferences for expansion. Some scholars suggest that international bureaucrats will always seek to increase their organization's scope. International organizations (or agents) have fixed interests to maximize their budget, tasks, and autonomy, as they are "competence-maximizers."[37] Staff will favor expansion if it enables them greater autonomy. Other scholars have argued that international organizations behave like firms and pursue financing, thus their expansion is contingent on the presence of external material resources.[38] Expansion here is a logic for survival given that international organizations operate in an increasingly competitive marketplace. According to this view, espoused in resource dependency theory, we would expect international organizations to expand if there were resources available to do so.

In addition, scholars have suggested that international bureaucrats may base their decisions on their perception of member state support. Staff are likely to be very aware of their dependency on states and will avoid advocating policies or programs that might jeopardize their relationship with states. They may "restrain themselves and advocate preferences in line with their principals if they think it will keep their organization alive."[39] Equally, they may be wary of behavior that might undermine the legitimacy of their organization's mission.[40] This account of international organizations suggests that staff's perception of whether states will support it is an important determining factor of expansion. Resource flows, organizational autonomy and state preferences may also factor in bureaucrats' decisions to expand or not. However, this book stresses the primary importance of how international bureaucrats view new issues and their place within existing organizational mandates. In particular, it identifies how new issue-linkages emerge, and are constructed and interpreted by international bureaucrats. If staff see the issue as having a strong issue-linkage to their core mandate they are likely to engage, but if not they are unlikely to expand. This is an important contribution to our understanding of change in international organizations.

Studying mandate change

Mandates outline the responsibilities of an international organization, and are set by states. The founding documents—typically a charter, constitution, or statute—describe the core functions and purpose of an organization. The challenge for scholars is that mandates can and do evolve over time, and their parameters are somewhat ambiguous.[41] In this book I distinguish between the formal delegated mandate from member states and the scope of activities that an international organization performs, regardless of member states' support. To identify the formal mandate we need to look at the declarations of member states at the institution's decision-making body—the council, executive committee, or equivalent. If member states agree collectively to add a new task or issue to an organization's mandate then this constitutes mandate change.

It is not always easy to identify mandate change. An organization may expand autonomously into a new issue-area and states may not express any opposition or reprimand the organization. This occurs as member states rarely monitor every action, program, or policy of an international organization. However, it makes it hard to know if states are granting tacit consent, or whether they are simply unaware of mandate creep but would be opposed to it. To identify mandate change I look for evidence that states have been informed of an action at the organization's decision-making body (council, executive board, or equivalent) and have collectively agreed to support it. This is not the same as a few member states endorsing and financing an organization to perform a new task without support from the entire council. This model of individually contracting to an institution is becoming increasingly common with the bilateralization of many multilateral agencies.[42]

To examine if and how mandate change is occurring across three organizations I firstly examine the international organizations' original mandates. I also examine the evolution of each institution between its inception and 1990, and focus specifically on whether the organization had a mandate for working on environmental issues in the 1990–2000 period. The 1990s saw intense international attention on the environment—partly because of the 1992 Rio United Nations Conference on Environment and Development (UNCED)—and new issue-linkages emerged between the environment and development, displacement, and migration. I show how member states had not mandated any of the organizations to work on climate change. This sets a baseline for analyzing mandate change in the 2000s.

The substantive sections of each case study chapter trace the rhetorical, policy, structural, and operational changes that UNHCR, IOM, and UNDP made in response to climate change between 2000 and 2014. I look at these four dimensions of change in each organization, as mandate change is not just about what happens in the top decision-making body, but also what happens on the ground and in speeches, policies, and staffing decisions. In other words, it is a gradual, iterative process that is influenced by international bureaucrats. Each of the four dimensions highlights a different type of change.

Rhetorical change occurs when executive heads and/or senior management prioritize climate change in their speeches. This dimension charts how the leadership presents their organization's position on a new issue-area. Policy change occurs when an organization develops and publishes new policy statements specifically on an issue. It typically demonstrates a more in-depth engagement with an issue than rhetorical change, and can come from any department of the bureaucracy. Structural change indicates that an international institution has established new departments or positions to focus on climate change. This requires significant resourcing and indicates a more long-term commitment to working on an issue than a single policy statement or speech. Finally, operational change occurs when an organization develops and implements new projects or programs with a focus on climate change. It refers to the actual services that an international organization delivers and focuses on how statements, policies, and structural change at headquarters actually translate to action on the ground.

These four categories are useful to identify, analyze, and structure our analysis of organizational change, and could be used for other issues, or other organizations. The case study chapters examine the changes along these four dimensions and if and how they led to mandate change. They are loosely chronological, and use process tracing, to examine rhetoric, policy, structure, and operational changes. However they do not follow a blow-by-blow account of each change and on some occasions the changes are grouped together, as they are intertwined. These four dimensions can also be reinforcing: if structural change occurs it is more likely we will see other dimensions of change as staff are hired to pursue an issue.

Research design

So why focus on only these three organizations, UNHCR, UNDP, and IOM? Many social scientists would argue it would be best to study a much larger group to increase the generalizability of theorizing about organizational change. This study focuses on only three organizations because it aims to examine in-depth how organizations engage with a new issue-area, for which there is no preexisting scholarship. Furthermore, the data on international organizations' engagement with climate change does not easily lend itself to quantitative analysis. An approach some scholars might advocate is a quantitative study analyzing the proportion of an institution's budget dedicated to climate change initiatives. Such studies have been completed for the World Bank, which has a specific environmental budget,[43] although they have not been without critics.[44] However this approach would not work with other agencies as usually they have no specific budget allocation, and it is extremely difficult to identify what exactly a climate-related activity or initiative is. In many cases international organizations during the course of this study were still devising methods to separate out climate change activities from other existing work. For instance, the Organisation for Economic Co-operation and Development (OECD), which tracks official development assistance, only began monitoring climate adaptation in 2010 and is still refining its measures. This study enriches our understanding of what counts as climate adaptation by looking at how international organizations have developed their work in this area.

Another alternative approach would be to study the evolving discourse of these international organizations and analyze the frequency, timing, and framing of climate change in their institutional texts. This approach has been followed by other scholars, in particular highlighting the security framing that many actors have used to link climate change to displacement.[45] However, a limitation of this approach is that it does not tell us if and how these discourses actually shape the practice of an international organization. Moreover, they tend to focus more on institutional headquarters, which issue most statements, speeches, and policy documents. Instead, this book traces organizational change from headquarters to operational activities on the ground. It seeks to provide a cross section of how international organizations engage with new issues and for this reason focuses on rhetoric, policy, structure, and operations. To complete such an in-depth analysis this book can focus on only three organizations.

For comparative purposes these organizations share a number of important similarities. Most importantly for the purposes of theorizing they are all intergovernmental organizations. This means they are created, funded, and mandated by states to perform particular tasks on their behalf. This is in marked contrast with NGOs, or businesses, which are private institutions, although they may act in the public good and have the autonomy to pursue their own agendas.[46] Although they vary in terms of voting procedures, regularity of board meetings, and their budgets, they are united in being organizations that operate for and on behalf of states, and several schools of international relations literature have been dedicated to explaining their actions.

Moreover, all three are international agencies with sizeable bureaucracies of between 6,500 and 8,500 staff. They are all served by international civil servants who must be loyal to their organization's mission rather than a particular national interest. They differ from other purely "forum" international organizations which are primarily to convene member states—think of the United Nations General Assembly, or various regional organizations such as the Organization for American States (OAS). They are not international secretariats, which assist states in the negotiation of international agreements, such as the United Nations Framework Convention on Climate Change (UNFCCC) secretariat. Rather these three organizations all deliver services and have field offices in over one hundred countries around the world. The organizations examined here are thus united by the fact that they have a large body of international staff, which deliver projects in developing countries, and are based in the Global North.

Finally, all were created in the post–World War II era: UNHCR and IOM in 1951, and UNDP in 1965. Thus a similar length of time (approximately 60 years) has elapsed between their original mandates and the present day. This is important as the book examines how they have interpreted their evolving mandates. However, these three organizations are not a representative sample of the total population of international organizations.[47] They were selected because each of them was established to focus on a single issue-area (development, refugees, and migration) and not to work on climate change.

These organizations do differ in significant ways. UNDP is a UN program based in New York, UNHCR is a commissioner based in Geneva; and the IOM is outside the UN system (although headquartered in Geneva). They vary also in scope: UNDP has a broader, and thus more flexible mandate (development) than UNHCR (refugees). They also differ in terms of functions: UNDP and IOM are service-orientated organizations delivering projects, while UNHCR also offers states a forum to negotiate new agreements and norms on international protection for refugees, stateless, and other displaced peoples. These differences will be discussed in subsequent chapters.

Sources and methods

To analyze organizational change at the rhetoric, policy, structural, and operational levels this book draws on a wide range of primary and secondary sources. I have built on the secondary literature of each organization's history and evolution, which is detailed in the case studies. I also analyzed all of the publicly available speeches, reports, policy papers, and executive committee proceedings relating to the issue of climate change in each of the three organizations. Analyzing these documents involved 1) identifying documents that mentioned climate change; 2) examining if and how the organization framed and justified its climate change work; 3) examining member states' support for climate change actions (in the case of executive reports); and 4) evaluating whether the document represented a qualitative change from previous statements, policies, and reports. The documents varied dramatically in type and scale: from four-page policy papers to one-hundred-fifty-page reports. It would have been problematic to compare them for the frequency of times they mentioned climate change or the strength of their justification for climate change action. There was thus no universal coding scheme used as it would have blurred important differences between reports rather than enabling the identification of the core documents and speeches in which rhetoric and policy evolved.

Secondly, I conducted semistructured interviews with over 150 officials from international organizations, member-states, and NGOs. These were conducted at: the UNFCCC negotiations in Copenhagen (2009); the UNHCR and IOM headquarters in Geneva (March 2010); the UNDP headquarters (October 2010); field operations in Kenya (March–April 2011); and the Nansen Conference on Climate Change and Displacement in Oslo (June 2011). Interviewing and observing international organizations at these various sites offered a much broader view of how they are adapting beyond their headquarters. The quality and quantity of these interviews offers in-depth insights into the decisions made and how these translated into action around the globe. Almost all interview participants are kept anonymous, in keeping with their wishes.

Kenya is a good site to analyze how development, migration, and refugee organizations are responding to climate change as it is one of the world's most drought-prone countries. In recent years drought has dried out pastures in northern Kenya, resulting in the deaths of cattle and devastating the primary livelihoods of Kenya's nomadic pastoralists. It also led to famine in Somalia, and thousands fled into Kenya to escape malnutrition, persecution, and ongoing conflict. In August 2009

Somalis were arriving into Dadaab, the world's largest refugee camp, at an "overwhelming average rate of 6,400 a month."[48] The president of Kenya, Mwai Kibaki, declared the drought and the resulting famine a national disaster.[49] The government explained that "10 million people are food insecure and require emergency support. These people will not be able to meet their minimum food requirements [without emergency measures]."[50] President Kibaki called on the international community to assist Kenya and launched an appeal for US$400 million in foreign aid. Drought and famine are not new to the Horn of Africa. However, the 2009 drought did exemplify the International Panel on Climate Change's (IPCC) prediction of a world with more severe and frequent extreme weather events.[51]

It also represents a "hard case" for UN agencies to justify expansion into environmental and climate change activities. Kenya is the global headquarters of the United Nations Environment Programme (UNEP) and it has a large operational presence there. UNEP carries out environmental activities in partnership with, and on behalf of, other international organizations. Thus we might expect non-environmental organizations to partner with UNEP rather than expanding and developing new expertise in this area themselves. Yet all three case study organizations have established climate change adaptation projects in Kenya and showcased these at the UNFCCC summit in Copenhagen. By focusing on Kenya we can examine if and how organizations are beginning to respond to climate change on the ground.

I identified interview participants in Kenya, Geneva, New York, and elsewhere according to their role in the three case study international organizations. I sought to interview those who worked specifically on climate change initiatives as well as senior managers with an overview of the organization. I also interviewed diplomats that were involved in executive committee deliberations, in monitoring and/or funding international organizations. I talked with a range of staff from partner organizations—international organizations and NGOs—who worked with the case study international organizations on climate change. Participants were found through a combination of the "snowball" sampling and directly contacting embassies and organizations for interviews. I also had regular telephone and in-person interviews with government and UN officials to keep abreast of policy developments and gain insights into operational change globally. These interviews helped to identify key changes in rhetoric, policy, structure, and operations; to contextualize primary and secondary literature; and to provide the institutional backstory of if and how these organizations pursued mandate change.

There were some limitations in this data collection. I was not able to formally interview current or previous heads of UNHCR, UNDP, or IOM, except for James Speth (administrator of UNDP, 1993–1999). My analysis of these leaders' motivations and perceptions is based on discussions with staff that have had close interaction with them. Interviews were temporally biased towards the 2009 to 2013 period when I conducted the bulk of interviews. I addressed this bias by talking with former staff from each international organization to gain insights into earlier periods. My analysis of member state and international organization relations for these earlier periods draws on executive committee reports and documents available online and the institutional memory of staff present.

To deepen my analysis of organizational change I also conducted participant observation at the various field site locations.[52] Participant observation is a process of collecting data by participating in and observing everyday life from the inside and is rarely used in international relations.[53] I used participant observation as a method of data generation and contextualization rather than seeking to create full organizational ethnographies. At the UNFCCC I observed where, how, and to whom each organization presented its climate change agenda. In Kenya I spent ten days based in Kakuma and two days in the Dadaab refugee camps and several weeks in Nairobi in 2011. I visited project sites in the refugee camps and in the surrounding areas, talked and shared meals and leisure time with IOM and UNHCR staff and, to a lesser extent, with beneficiaries in the refugee and host community. This work was crucial to understanding if and how international organization operations were changing on the ground. I sought rich detailed accounts of international organization change that went beyond formal interviews with bureaucrats and policy documentation, which are often the dominant data sources for international relations scholars of international organizations.

Structure of the book

This subsequent chapter traces how climate change became linked to development, migration, and displacement. It begins by looking at the emergence of adaptation in the UNFCCC negotiations in the 2000s. The broadening of the focus from mitigation to adaptation enabled a host of new actors to forge issue-linkages with climate change. The chapter's core contribution is to examine how civil society, academics, and epistemic communities established links between climate change and migration, and climate change and development. It examines the evolution of these issue-linkages which are now often taken for granted yet form the rationale for development, migration, and refugee organizations' engagement with climate change. The subsequent three chapters then explore how and why UNHCR, IOM, and UNDP integrated climate change into their rhetoric, policy, structure, and operations. Each chapter examines the evolving mandate of the organization, before concentrating on its engagement with climate change (2000–2015).

The third chapter traces how High Commissioner António Guterres sought a role for UNHCR in climate change and displacement. It contrasts this with the resistance of many staff and states that were concerned engagement with climate displacement would undermine the office's role in refugee protection. It demonstrates how policy and rhetoric on climate change took off at the global level in 2008–2010 but not in operations on the ground in Kenya, despite severe drought. The chapter emphasizes that UNHCR was initially resistant to engaging with climate change, but then from 2010 onwards sought to carve out a new protection role and build on its existing protection role for refugees and on humanitarian work with internally displaced peoples.

The subsequent chapter examines how IOM engaged in the 1990s with the environment and migration, and in the mid-2000s this work evolved into climate change and migration. The office researched and wrote reports on the issue, advocated for an interagency standing committee on climate migration, and initiated 500 new climate change related projects, including in Kenya. It suggests that IOM staff pursued new financing opportunities as they saw no conflict with IOM's core mandate, even though it had little expertise in adaptation.

The fifth chapter finds that UNDP expanded into environmental issues in the 1990s through its work on sustainable human development, and with multilateral financing from the Global Environment Facility (GEF). In the mid-2000s it became one of the main agencies in the UN system developing climate change adaptation policies and operations, with a large portfolio backed by a substantial grant from the Japanese government and multilateral financing. UNDP staff also highlighted the impacts of climate change for development in various Human Development Reports. Most UNDP staff saw no conflict between UNDP's development mandate and work on climate adaptation, and drove expansion. However, it was only in the late 2000s that member states granted UNDP an explicit mandate to work on climate change adaptation.

The conclusion highlights the role of staff in driving changes when they saw clear, strong issue-linkages between their mandates and climate change. It also shows that states set some important parameters for organizational change and expansion. Finally, it draws out the broader implications of this work for global governance. It suggests that climate change governance is much broader than commonly conceptualized, and scholars should investigate how a range of institutions which are not obviously environmental have engaged with the regime. It encourages scholars to examine if and when organizations are effective when working in new issue-areas. This book has important policy implications as we look to these global institutions to resolve twenty-first century problems.

Conclusion

This book focuses on three institutions: the UNHCR, the UNDP, and IOM. All three were designed in the post–World War II era in response to the devastating effects of war, and none was given a mandate for climate change or environmental issues. This book traces their responses to climate change in their rhetoric, policy, structure, operations, and overall mandate change. This book suggests that international bureaucrats can play a decisive role in deciding whether and how to expand into a new issue-area. The next five chapters make a critical contribution by illustrating how international organizations forge, frame, and internalize new issue-linkages and use these to advocate for expansion. The book helps us to understand how institutions established in the twentieth century are adapting to a twenty-first century world.

Notes

1 Vikram Kolmannskog, "Climate Change, Human Mobility and Protection: Initial Evidence from Africa," *Refugee Studies Quarterly* 29, no. 3 (2010): 103–119; Norman Myers, "Environmental Refugees in a Globally Warmed World," *BioScience* 43, no. 11 (1993): 752–761.
2 Elizabeth L. Malone, *Debating Climate Change* (London: Earthscan, 2009).
3 John Barnett, "Adapting to Climate Change in Pacific Island Countries: The Problem of Uncertainty," *World Development* 29, no. 6 (2001): 933–977.
4 Myers, "Environmental Refugees in a Globally Warmed World."
5 Frank Biermann and Ingrid Boas, "Preparing for a Warmer World: Towards a Global Governance System to Protect Climate Refugees," *Global Environmental Politics* 10, no. 1 (2010): 60–88.
6 Robert Keohane and David Victor, "The Regime Complex for Climate Change," *Perspectives on Politics* 9, no. 1 (2011): 7–23.

7 Sikina Jinnah, "Climate Change Bandwagoning: The Impacts of Strategic Linkages on Regime Design, Maintenance, and Death," *Global Environmental Politics* 11, no. 3 (2011): 1–9.

8 Sikina Jinnah, *Post-Treaty Politics, Secretariat Influence in Global Environmental Governance* (Cambridge, Mass.: The MIT Press, 2014).

9 Joseph Jupille, Walter Mattli, and Duncan Snidal, *Institutional Choice and Global Commerce* (Cambridge, Mass.: Cambridge University Press, 2013).

10 Ibid.

11 Tana Johnson, *Organizational Progeny* (Oxford: Oxford University Press, 2014).

12 Jacqueline Best, "Ambiguity and Uncertainty in International Organizations: A History of Debating IMF Conditionality," *International Studies Quarterly* 56 (2012): 674–688.

13 Darren G. Hawkins, David A. Lake, Daniel L. Nielson and Michael Tierney, "Delegation under Anarchy: States, International Organizations, and Principal-Agent Theory," in *Delegation and Agency in International Organizations*, ed. Darren G. Hawkins et al. (Cambridge, Mass.: Cambridge University Press, 2006), 3-38.

14 Ibid.; Mark A. Pollack, *The Engines of EU Integration: Delegation, Agency and Agenda Setting in the EU* (Oxford: Oxford University Press, 2003).

15 Hawkins et al.,"Delegation under Anarchy: States, International Organizations, and Principal-Agent Theory."

16 Pollack, *The Engines of EU Integration: Delegation, Agency and Agenda Setting in the EU.*

17 "Police patrols" involve active monitoring of international organizations' behavior with the aim of detecting violations of legislated goals. Meanwhile "fire alarms" rely on third parties (such as interest groups and citizens) to monitor agencies. Ibid., 42; Daniel L. Nielson and Michael J. Tierney, "Delegation to International Organizations: Agency Theory and World Bank Environmental Reform," *International Organization* 57, Spring (2003): 241–276.

18 Susan Park and Catherine Weaver, "The Anatomy of Autonomy," in *International Organizations as Self-Directed Actors*, ed. Joel E. Oestrich (Abingdon: Routledge, 2012), 91–117.

19 Inis Claude, *Swords into Plowshares* (New York: Random House, 1956); Michael N. Barnett and Martha Finnemore, *Rules for the World, International Organizations in Global Politics* (Ithaca, NY: Cornell University Press, 2004).

20 Alexander Betts, "Substantive Issue-Linkage and the Politics of Migration," in *Arguing Global Governance*, ed. Corenliu Bjola and Markus Kornprobst (Abingdon: Routledge, 2010), 85–100.

21 Laura Shephard, *Gender, Violence and Security: Discourse as Practice* (London: Zed Books, 2008).

22 Daniel Esty, "Bridging the Trade-Environment Divide," *Journal of Economic Perspectives* 15, no. 3 (2001): 113–130.

23 Betts,"Substantive Issue-Linkage and the Politics of Migration,"; Ernst Haas, "Why Collaborate? Issue-Linkage and International Regimes," *World Politics* 32, no. 3 (1980); Oran Young, "Institutional Linkages in International Society: Polar Perspectives," *Global governance* 2 (1996): 1–24.

24 Betts, "Substantive Issue-Linkage and the Politics of Migration," 2.

25 There is a potential problem of endogeneity with this conception of issue-linkage. International organizations could create and promote an issue-linkage between their regime and another and thus contribute directly to changing linkage salience. They could then use the strong issue-linkage as a basis for expansion. A hypothetical example: UNEP could write an influential report that demonstrated how international trade impacted on the environment. This report could lead other actors (NGOs, states and other international organizations) to support a trade-environment linkage. Thus, UNEP would have effectively increased the linkage salience of trade to the environmental regime. UNEP may then cite this issue-linkage (and the report) as a basis for involvement in WTO negotiations. Endogeneity would be an issue if a single organization or report could change relative salience between two regimes. However, I seek to separate out the wider context of an emerging issue-linkage from a single organization's attempts to create their own institutional issue-linkage. A single organizational report or speech is not sufficient to change the overall strength of an issue-linkage: the linkage is intersubjective. Thus there must be a number of actors creating, advocating for, and maintaining the issue-linkage.

26 David Mitrany, *The Functional Theory of Politics* (London: London School of Economics and Political Science, 1975).

27 Joseph Samuel Nye, "Comparing Common Markets: A Revised Neo Functionalist Model," in *International Organization, a Reader*, ed. Friedrich Kratochwil and Edward D. Mansfield (New York: Harper Collins College Publishers, 1994), 286–301.

28 Bob Reinalda and Bertjan Verbeek (eds), *Autonomous Policy Making by International Organizations* (London: Routledge, 1998).

29 Andrew P. Cortell and Susan Peterson, "Dutiful Agents, Rogue Actors, or Both? Staffing, Voting Rules, and Slack in the WHO and WTO," in *Delegation and Agency in International Organizations*, ed. Darren G. Hawkins et al. (Cambridge, Mass.: Cambridge University Press, 2006), 255–280.

30 Cortell and Peterson, "Dutiful Agents, Rogue Actors, or Both? Staffing, Voting Rules, and Slack in the WHO and WTO," 260.

31 Barnett and Finnemore, *Rules for the World, International Organizations in Global Politics*, 44.

32 Ibid., 44.

33 Ibid., 64–65.

34 James March and Johan P. Olsen, *Rediscovering Institutions, the Organizational Basis of Politics* (Ontario: The Free Press, 1989), 22.

35 Ibid.

36 Michael N. Barnett and Liv Coleman, "Designing Police: Interpol and the Study of Change in International Organizations," *International Studies Quaterly* 49, no. 4 (2005): 593–620.

37 Pollack, *Engines of EU Integration: Delegation, Agency and Agenda Setting in the EU*, 39.

38 Erica R. Gould, "Money Talks: Supplementary Financiers and International Monetary Fund Conditionality," *International Organization* 57, no. 3 (2003): 551–586.

39 Cortell and Peterson, "Dutiful Agents, Rogue Actors, or Both? Staffing, Voting Rules, and Slack in the WHO and WTO," 261.

40 Cortell and Peterson, "Dutiful Agents, Rogue Actors, or Both? Staffing, Voting Rules, and Slack in the WHO and WTO," 267.
41 Best, "Ambiguity and Uncertainty in International Organizations: A History of Debating IMF Conditionality."
42 Erin Graham, "Money and Multilateralism: How Funding Rules Constitute IO Governance," *International Theory* 7, no. 1 (2015): 162–194.
43 Nielson and Tierney, "Delegation to International Organizations: Agency Theory and World Bank Environmental Reform."
44 Tamar Gutner, "World Bank Environmental Reform: Revisiting Lessons from Agency Theory," *International Organization* 59 (2005): 773–783.
45 Chris Methmann and Angela Oels, "From 'Fearing' to 'Empowering' Climate Refugees: Governing Climate Induced Migration in the Name of Resilience," *Security Dialogue* 46, no. 1 (2015): 51–68.
46 Volker Rittberger and Bernhard Zangl, *International Organization: Polity, Politics and Policies* (Basingstoke: Palgrave Macmillan, 2006).
47 International organizations encompass a range of different organizations from regional groupings (the Organization of American States, the Arab League, the Pacific Regional Environment Programme, the Association of Southeast Asian Nations [ASEAN], or the Organization of Petroleum Exporting Countries [OPEC]) some of which are highly integrated, such as the EU, and others which are less so, such as the Economic Community of West African States (ECOWAS). Intergovernmental organizations also encompass: international courts (the International Criminal Court; the EU Court of Justice), regional development banks (the African Development Bank; the European Bank for Reconstruction and Development, the Asian Development Bank), and other issue-specific organizations (the EU Council of Ministers of Transport, the International Institute of Refrigeration, the International Tropical Timber Organization). In addition, there are entities affiliated with the UN: from the Bretton-Woods institutions (International Monetary Fund [IMF] and World Bank) to specialized agencies (World Health Organization [WHO], International Labour Organization [ILO], United Nations Children's Fund [UNICEF], the United Nations Universities) which have a higher degree of autonomy than the UN programs or funds. There are also UN secretariats which assist, and even direct, international negotiations—from the UNFCCC to the UN Convention on Biological Diversity (UNCBD).
48 This posed a major humanitarian challenge as the camp's "facilities and resources were already overstretched and hosted three times the population it was designed for." UNHCR 2009, *Conflict and Drought Force More Somalis to Leave Kenya*, www.unhcr.org/4abc9f829.html.
49 "Kenya seeks Sh32 bn food aid", *Daily Nation*, January 16, 2009, www.nation.co.ke/News/-/1056/515550/-/u1aefp/-/index.html.
50 Ibid.
51 United Nations Environment Programme (UNEP) claimed that the drought across East Africa was "worsening because of global climate change." Note that climate change increases the likelihood of extreme weather events, such as droughts, occurring but is not necessarily the sole cause. UNEP cited in: Mohammed Adow, *Kenya Drought "Has Spared No One,"* Al Jazeera, 2009. www.aljazeera.com/focus/2009/10/20091011941816643.html.

52 Hugh Gusterson, "Ethnographic Research," in *Qualitative Methods in International Relations*, ed. Audie Klotz and Deepa Prakash (Basingstoke: Palgrave Macmillan, 2008), 93–113.

53 Wanda Vrasti, "The Strange Case of Ethnography and International Relations," *Millennium: Journal of International Studies* 37, no. 2 (2008): 279–301.

1 Issue-linkages

- **Adaptation and the climate change regime**
- **Issue-linkages to climate change**
- **Conclusion**

The climate change regime is a regime "complex" as it was founded on a number of distinct cooperation problems.[1] These core issues include: the coordination of emission regulations; compensation for developing countries by developed countries for past emissions; the coordination of efforts to prepare for climate change, in particular adaptation and geoengineering; and the coordination of common scientific assessments. Although all of these issues were present at the inception of the climate regime, it is only in the last decade that adaptation has become a major issue of concern for states in the UNFCCC negotiations. This chapter traces the evolution of adaptation within the UNFCCC, the institution at the heart of the climate change regime. It does not cover debates over emissions reductions; for this the reader should look elsewhere.[2] Rather it contributes to a nascent literature on adaptation governance within the climate regime.[3]

The chapter focuses on the emergence of two substantive issue-linkages: climate change and development; and climate change and migration. It traces the work of civil society, academics, and epistemic communities in establishing causal connections between these distinct areas. These substantive issue-linkages offered a rationale for UNHCR, UNDP, and IOM to engage with climate change. These linkages shaped when and why these three organizations engaged with climate change and are critical contextual information for the following three case study chapters.

The chapter begins by examining when and how adaptation became a more central part of the UNFCCC negotiations. It suggests that the climate regime broadened with a more explicit focus on adaptation, as donor states pledged funds to assist developing countries to deal with and prepare for the impacts of climate change, and new financing mechanisms were established. This opened the door for a number of international organizations with no environmental mandate to directly engage in the climate change negotiations. It then unpacks how issue-linkages between climate change and development and climate change and migration were forged. This is important as today it may seem obvious to many that climate change will have an impact on developing countries, and lead to migration and displacement. However these linkages were not so evident in the early 1990s, and the exact nature of the causal connection between climate change and migration is still debated today. In fact, this chapter will also be of interest to scholars of issue-linkages as it examines how such linkages evolve and strengthen.

Adaptation and the climate change regime

In the 1990s and early 2000s the dominant focus of the UNFCCC, as well as scientists and activists, was to stop climate change through mitigation, reductions in greenhouse emissions.[4] Adaptation (preparing for and dealing with the impacts of climate change) was not a focus for the IPCC or the UNFCCC.[5] Many saw adaptation as a dangerous agenda item. At best, focusing on adaptation might distract states' attention and resources away from mitigation. At worst, it represented defeat for states seeking emissions reductions as they would have to acknowledge that mitigation was not sufficient to combat climate change.[6] Environmentalists and scientists were also wary of discussing adaptation. Although both the UNFCCC agreement and the first IPCC report (1990) mentioned adaptation, it was a marginal issue for both that received little attention.[7] In fact, during the first ten years the IPCC did not devote a single chapter of their reports to adaptation to climate change.[8]

The emergence of adaptation

In 2000 adaptation began to gain traction when the IPCC held their first workshop explicitly on adaptation. This workshop became Working Group II on "Impacts, Adaptation and Vulnerability," one of three working groups of the IPCC. The group wrote its first chapter on adaptation in the IPCC's Third Annual Assessment in 2001 and directed states to respond to the impacts of climate change.[9] Adaptation also began to be a more central part of the UNFCCC. As the negotiations over the Kyoto Protocol became difficult, the EU agreed to establish an annual climate change fund of US$15 million to target adaptation as well as mitigation. Meanwhile, the new US president, George W. Bush, announced he would withdraw the United States from the Kyoto Protocol, which President Bill Clinton had previously signed.[10] The absence of the world's largest economy and emitter jeopardized an agreement; however other states continued to negotiate and in 2001 finalized the Kyoto Protocol.

At the 2001 Conference of the Parties (COP) 7 of the UNFCCC in Marrakech, states began to prioritize the adaptation needs of Least Developed Countries (LDCs). In the final agreement states established the Least Developed Countries work program and called for LDCs to identify and report their adaptation needs through the National Adaptation Programmes of Action (NAPAs).[11] NAPAs became a key reporting and policy-making tool for adaptation in many developing states. They established the Least Developed Countries Fund to assist LDCs to develop these National Adaptation Programmes for Action. In addition, states also established two other multilateral funds: the Special Climate Change Fund, based on voluntary donations from states to facilitate technology transfer from developed to developing states; and the Adaptation Fund which was financed by a two percent levy on the Clean Development Mechanism. The establishment of these three climate funds signaled that adaptation had become a necessary response to climate change.

Adaptation was institutionalized as a top priority, on par with mitigation, between 2006 and 2014. The "African" COP in Nairobi in 2006 pushed this change. African states saw adaptation as a top priority and argued that they were amongst the most vulnerable to the impacts of climate change.[12] At Nairobi states finalized the Nairobi Work Programme on adaptation, a five-year plan of work to support climate adaptation which sought to enhance information exchange and capacity to identify vulnerability and implement effective adaptation responses.[13] At Nairobi they also officially launched the Adaptation Fund; states had spent the previous six years elaborating its governance and operational structure. Adaptation became a politicized developing-country issue, and not just a technical concept linked to discussions over impact thresholds. Developing states and their allies argued that those who suffer most are the least responsible for climate change.

Then at COP 13 in 2007 states initiated the Bali Action Plan, which called for "enhanced action on adaptation."[14] In particular it suggested the consideration of "the urgent and immediate needs of developing countries that are particularly vulnerable to the adverse effects of climate change." The Bali Action Plan set out the framework and priorities for the two-year lead-up to the negotiations in Copenhagen, which were billed as the critical negotiations to determine what would follow on after the Kyoto Protocol. Adaptation also featured strongly in the deliberations of the Ad Hoc Working Group on Long-term Cooperative Action under the Convention (AWG-LCA).[15]

The 2007 IPCC fourth report reinforced the call for adaptation measures: no longer was climate change a future problem but it was here now. Global warming was predicted to have adverse effects on everything from rural livelihoods (crops may no longer grow in some areas as rain frequency and temperatures change) to the spread of disease (malaria mosquitoes will inhabit a much larger area).[16] Increased extreme weather events from floods to droughts could undermine development programs and be an obstacle to achievement of the Millennium Development Goals.[17] The IPCC report emphasized that developing states were the most "vulnerable" to climate change due to their lack of financial or technological resources. A comparison of the Netherlands and Bangladesh, two low-lying states which are largely dependent on agricultural production, illustrates this. The Netherlands has a much greater capacity than Bangladesh to adapt because of its economic resources and technical skills.[18] Development agencies looked to include adaptation activities and "climate proof" their development activities.[19] Meanwhile civil society organizations, international organizations, and NGOs increased their pressure on states at the UNFCCC to address climate adaptation. Between 2007 and 2009 there was a dramatic increase in the number and range of international NGOs and international organizations attending the UNFCCC.[20]

At the UNFCCC negotiations in Poznan in 2008 discussions of binding emissions targets stalled, however delegates discussed new funding for mitigation and adaptation. States agreed on principles for a dedicated fund to assist developing states with the impacts of climate change. A crucial criterion was that this funding had to be additional to current development spending. Developing states argued that existing development assistance should not be diverted. They were supported by a broad civil society movement calling for "climate justice": the Global North had an obligation to pay for adaptation as its industrialization process had led to climate change. For this reason, climate financing had to be distinct from, and on top of existing development financing commitments.[21] In the lead-up to Copenhagen, activists mobilized around the world demanding a binding agreement and additional climate financing for developing countries.

The Copenhagen climate change negotiations were widely seen as a massive failure.[22] State parties reached no agreement on the post-Kyoto Protocol, and the final day of negotiations carried on throughout the night and into that Saturday without a consensus. In the end, the primary document emerging from the negotiations was a political agreement drafted by the heads of states from the United States, China, India, Brazil, and South Africa during a private meeting. The three-page document committed parties to "scaled up, new and additional, predictable and adequate funding" and established a new "Green Climate Fund."[23] States pledged US$30 billion annually in additional climate financing for the 2010–2012 period and outlined their intentions to mobilize US$100 billion through a range of sources by 2020.[24] These pledges at Copenhagen were a significant breakthrough: till then states had only been invited to make financial contributions on a voluntary basis and no targets had been set. However, the accord was not formally agreed to by all states, rather they were invited to take note of the final agreement. How states would meet these pledges, and through what mechanisms the financing would flow became the focus for the following UNFCCC negotiations.

There was a general sense of disillusionment and despair after the failure of Copenhagen to secure a binding, international agreement on climate change. Expectations were much lower the following year at negotiations in Cancún and states successfully reached a final agreement, adopted by all parties, which set out a new "Cancun Adaptation Framework." The Framework was the result of the previous three years of negotiations following the Bali Action Plan and states agreed that "adaptation must be addressed with the same priority as mitigation and requires appropriate institutional arrangements to enhance adaptation action and support."[25] The agreement endorsed the COP pledges and requested that developed states provide developing countries "new and additional finance." Furthermore, it entrenched adaptation within the negotiations through the establishment of a new Adaptation Committee to promote enhanced action on adaptation under the convention and further elaborated the Green Climate Fund.

Adaptation remained a central topic of negotiations at Durban (2011), Doha (2012), and Warsaw (2013). At Durban states sought to develop the architecture for the new climate fund, in the midst of wider debates about what would happen when the Kyoto Protocol period ended. They elaborated the governance structure of the Green Climate Fund and in the final agreement specifically requested the board to balance the allocation of Green Climate Fund resources between adaptation and mitigation.[26] They also agreed on the composition of, as well as modalities and procedures for the Adaptation Committee to ensure adaptation work continued in the UNFCCC. The final agreement also noted the need to support developing countries to develop National Adaptation Plans and a Work Programme to Consider Approaches to Address Loss and Damage.[27]

The 2012 Doha COP reiterated pledges to "enhance action on adaptation." The Copenhagen "fast track financing period" over the 2009–2012 period was ending and the final agreement urged developed states to "increase their efforts" to provide financing during the 2013–2015 period.[28] The agreement suggested the Adaptation Committee establish an annual adaptation forum in conjunction with the UNFCCC annual summit to keep adaptation high on the agenda. States also advanced the details of how the Green Climate Fund would operate. However, negotiations over the exact target of financing became tense in Warsaw in 2013. The Group of 77 (G77), a coalition of developing countries, and China led a walk-out during talks about "loss and damage" compensation and demanded that the developed world give them US$100 billion annually by 2020, as had been promised at Copenhagen. However, developed states resisted calls to set targets for climate financing for the rest of the decade. The final agreement of the conference merely mentioned setting "increasing levels" of aid.

Adaptation financing

While the UNFCCC negotiations saw the establishment of a number of new adaptation funds, as described above, financing also became available through other multilateral, bilateral and private mechanisms. In fact, mapping out all of the funds available for adaptation activities is extremely difficult given the plethora of sources.[29] To briefly summarize the scene today: there are five multilateral funds that focus exclusively on adaptation. These are the Special Climate Change Fund (administered by the GEF), the Least Developed Countries Fund (GEF), the Strategic Priority for Adaptation (GEF), the Pilot Program for Climate Resilience (World Bank), and the Adaptation Fund (administered by the Adaptation Fund Board). In total, by 2011 states had pledged US$1.34 billion through these five funds, however only US$581.2 million have been received thus far.[30] Both sums fall far short of the US$27–66 billion that the US estimates will be needed per year to help developing countries adapt.[31]

There are numerous other funds provided by international organizations that can be used for adaptation. These include: the African

Development Bank Climate Risk Management and Adaptation Strategy (CRMA), the Global Climate Change Alliance and the Asian Development Bank (ADB) small grants for adaptation actions targeted at Asian and Pacific states.[32] In addition, there are bilateral funds established by donor states to meet their Copenhagen pledges. In fact, a number of aid agencies, such as the United Kingdom, Germany, Norway, and Japan have established new climate change departments to manage these finances.

Some attempts have been made to track climate adaptation assistance, such as the OECD's Development Assistance Committee adaptation marker. Essentially, the OECD's Rio Markers for adaptation require donor states to code and report on the percentage of their projects that are "primarily" or "significantly" orientated to address adaptation. The OECD has agglomerated bilateral donors (but not multilaterals) and reported that adaptation aid reached US$9.3 billion on average per year in 2010–2012, representing 7.1 percent of all Overseas Development Assistance (ODA) commitments.[33] However, only US$2.7 billion targeted adaptation as a "principal objective" and the majority targeted multiple environmental objectives including mitigation, biodiversity, and desertification. In fact, the vast majority of climate financing goes to mitigation.[34]

These markers have been widely criticized as scholars have noted that states tend to overreport their climate change activities due to uncertainty about how to code them or a political desire to demonstrate high climate change spending.[35] Furthermore, tracking donors' climate change assistance is further complicated by the fact that adaptation and development activities are inherently intertwined, so it is difficult to disaggregate whether money spent on adaptation is distinct from existing development assistance. In fact, it is estimated that 60 percent of all ODA could intersect with adaptation activities.[36] Tracking how much "additional" financing has been directed to adaptation is near on impossible. This is because there is no accepted definition of what constitutes "additionality" of climate assistance.[37]

The next section now examines how the broadening of the climate regime to include adaptation opened the door for a range of non-environmental international organizations to become involved in the UNFCCC. There was a strong incentive to engage, given the predicted severity of climate change and its crosscutting nature, as well as the opportunity for new funding. There was also a gap in global governance: no international organization was tasked with a mandated responsibility for adaptation. UN agencies competed to be the "lead" adaptation agency at UN interagency meetings (namely the High Level

Committee on Programmes and the Chief Executives Board).[38] Secretary-General Ban Ki-moon decided not to nominate a lead agency and instead enable any UN agency to proclaim it could "do adaptation."[39] In fact, he wrote to most UN programs and agencies asking them to establish climate change task forces and elaborate their position on climate change, as will be discussed in the next chapter.

Issue-linkages to climate change

The broadening of the climate change regime to encompass adaptation enabled a range of new international organizations to engage with the UNFCCC, even though they were not environmental or climate change orientated, as the above section demonstrated. These actors made "substantive issue-linkages," elaborating how climate change impacted on their issue-areas and vice versa. This section now examines two specific substantive issue-linkages: climate change and development and climate change and migration, to set the scene for the subsequent case study chapters. It argues that a range of organizations and actors sought to establish these issue-linkages. It traces the evolution of each linkage from its emergence in policy, NGO, and academic documents to the presence of interagency groups focusing on the issue or senior politicians endorsing the link.

Climate change and development

The connection between development and climate change is embedded in the UNFCCC regime. Since the 1972 Stockholm conference global environmental issues and development have been intertwined. Developing countries have argued that environmental protection cannot undermine their development.[40] However, it is important to distinguish between three distinct issue-linkages: 1) development leads to climate change; 2) sustainable development can mitigate climate change; and 3) adaptation is sustainable development. These substantive issue-linkages emerged at different times, were endorsed by different actors, and have strongly influenced and framed discussions within the climate regime. It is the last substantive linkage that is most relevant as this book focuses on international organizations' adaptation initiatives, not mitigation activities.

The core argument underlying the UNFCCC is that development (in the form of industrialization) causes climate change. The issue-linkage is scientific: the shift from an agricultural based economy to one dependent on carbon-intensive manufacturing industries has led to a large increase in carbon emissions, which has led to an increase in global mean temperatures.[41] Although this has been contested by a number of "climate change deniers," we can state with a high degree of certainty (based on the IPCC reports) that the process of development and industrialization in the Global North has led to global warming.[42]

The notion that development has caused climate change framed the Kyoto Protocol and subsequent negotiations for a post-Kyoto framework. The Kyoto Protocol set out binding targets for developed states only, also known as Annex 1. Developing states did not have to make emissions reductions. The fundamental argument for this division in burden-sharing is that the industrialization of the Global North led to an increase in carbon emissions and resulted in anthropogenic global warming.[43] Meanwhile, most developing countries had not, or not until relatively recently, industrialized, and if so, to a much lesser extent. They also had extremely low carbon emissions and so did not have the same obligation to reduce their emissions.

The debate shifted in the mid-2000s as "emerging" economies such as China, India, and Brazil developed large industrial sectors and their carbon emissions increased.[44] The Global North, led in particular by the United States, argued that emerging economies should also commit to reducing their emissions in a future global agreement on mitigation. Meanwhile the coalition of G77 + China continued to argue that climate change was a result of industrialization over centuries in the Global North and thus the North should shoulder the largest responsibility for binding emissions cuts. In short, the causal connection between development and climate change is an issue-linkage that the UNFCCC negotiations are premised on.

If development can cause climate change, then *sustainable* development should mitigate it. Sustainable development entails reducing—or at least minimizing emissions—which we would otherwise see under "business as usual scenarios."[45] On this basis, developed and developing states should utilize renewable energy and promote sustainable industries and public transport. This issue-linkage was based on a new norm of sustainable development, coined in 1987 by the *Brundtland Report*, also known as the *Report of the World Commission on Environment and Development: Our Common Future*.[46] The report defined sustainable development as: "development that meets the needs of the present without compromising the ability of future generations to meet their own needs."[47] Essentially, development was seen to have had social and environmental costs, such as carbon emissions, pollution, and resource depletion. In contrast, sustainable development could promote economic growth that was not at the expense of the environment.

The 1992 United Nations Conference on Environment and Development in Rio de Janeiro elaborated the concept of sustainable development. The conference declared that "In order to achieve sustainable development, environmental protection shall constitute an integral part of the development process and cannot be considered in isolation from it."[48] In addition, at Rio states endorsed a new environmental fund, the Global Environment Facility (GEF) to finance environmental projects in developing countries.[49] The GEF focused on four key areas: biodiversity, climate change, international waters, and ozone depletion (inorganic pollutants). The World Bank, UNDP, and UNEP had drafted the GEF constitution and were the only agencies with access to its funds. The GEF led to a dramatic increase and expansion of environmental activities in the World Bank and UNDP.[50] Thus by the mid-1990s there was a strong issue-linkage between sustainable development and mitigation of climate change. Furthermore the World Bank and UNDP, the two largest development agencies, had exclusive access to a multilateral fund to mitigate climate change. UNDP's initial engagement with climate change mitigation is linked to the establishment of the GEF in the 1990s, and based on the issue-linkage between sustainable development and mitigation of climate change. This will be discussed in Chapter 5.

The focus of this book however is on climate change adaptation, not mitigation, which became more central to the UNFCCC negotiations in the 2000s. Development NGOs were central advocates of this shift and forged a substantive issue-linkage between adaptation and development. From the mid-2000s onwards development NGOs began to publish major reports on the need for developed states to assist developing countries to adapt to climate change. These NGOs were part of a global transnational movement demanding "climate justice" or compensation for the emissions released over the centuries by developed states. However, the rationale for development agencies to engage in climate change was not always obvious: climate change was perceived to affect polar bears, the focus for development agencies was people. The "Up in Smoke" coalition in the UK was in the vanguard of a shift. In 2003 the International Institute for Environment and Development (IIED) and the New Economics Foundation convened a series of discussions between environment and development agencies to work out the common ground between them.[51] From this they published a series of reports that were among the first to describe human experiences of climate change in the developing world.

These conversations and reports catalyzed a number of development agencies, such as Oxfam, to develop climate change policies and campaigns. Oxfam, for instance, published more than six reports on climate change annually from 2006 onwards. It also worked with Cooperative for Assistance and Relief Everywhere (CARE International), UNEP, the World Health Organization (WHO), and the International Federation of Red Cross and Red Crescent Societies (IFRC) to establish the Global Humanitarian Fund in 2007 to examine the impact of climate change on people.[52] They published a major report—*The Anatomy of a Silent Crisis*—immediately before the UNFCCC summit in 2009. Many donors from the United Kingdom to Norway also sought expertise on adaptation. Adaptation was widely recognized as an important issue by the leading development agencies and donors by 2007.

Between 1992 and 2014 there was also a major shift in how adaptation was conceptualized. During the 1990s, adaptation was primarily seen as a technical response to a specific climate change impact or vulnerability in a particular place.[53] It was defined in the first assessment report of the IPCC in 1990 as "measures to reduce the impact of global climate change," and primarily conceived as a technical activity such as engineering to protect against sea-level rise.[54] Over a decade later, in the fourth IPCC report (2007), adaptation was redefined more broadly to be any "adjustment in natural or human systems in response to actual or expected climatic stimuli or their effects, which moderates harm or exploits beneficial opportunities."[55] Today this broader understanding of adaptation dominates.

The malleability of adaptation facilitated the establishment of a wide range of new issue-linkages.[56] One representative of the Office for the Coordination of Humanitarian Affairs (OCHA) staff member even claimed that "No one is a specialist in adaptation, no one knows what it is."[57] Thus the term could be adapted and adopted to suit any interest. UNDP for instance promotes "pro-poor and pro-growth adaptation" which they define as "supporting countries to integrate climate-related risks and opportunities into national planning and poverty reduction, while addressing the needs of more vulnerable groups like women and indigenous people."[58] Meanwhile, the WHO sees adaptation as decreasing "health vulnerability to current climate variability and future climate change."[59] Adaptation activities could be a vaccination program to eradicate disease in low-income areas or building sea-walls in low-lying islands.[60] In short, adaptation encompasses everything from sustainable development to humanitarian relief to reproductive health rights, depending on which organization is defining it.

Alongside the policy debates, an academic literature emerged examining the connections between vulnerability, adaptation, and development, which also fed into the IPCC working group II discussions.[61] Individual vulnerability to climate change was predicated on factors such as class, ethnicity, age, and gender. Development targeted the marginalized and addressed vulnerabilities of the most disadvantaged and was thus closely interlinked with adaptation. Richard Klein, for instance, one of the leading academic experts on adaptation, established a new journal in 2009 entitled *Climate Change and Development* to investigate the linkage between climate change adaptation and development. In his inaugural introduction he emphasized that climate research and development studies have "their own separate history and tradition. Yet the connections between climate and development can no longer be ignored."[62]

Academic debate sought to clarify the relationship between adaptation and development. For some, "almost any development project could be reframed as an adaptation project" because adaptive capacity could be limited by any factor of development.[63] Others argued that there was a continuum from adaptation as development to "adaptation as technical solution to specific impact," which was distinct from development.[64] In the former, adaptation activities could include mainstream development work such as diversifying livelihoods. In the latter, adaptation was specific to confronting climate change impacts by reducing specific risks of flooding, drought or sea-level rise. These academic and policy debates about what constitutes adaptation and how we can distinguish it from development continue.

In summary, the linkage between adaptation and development has existed since the establishment of the UNFCCC in the 1990s. However the first decade of UNFCCC negotiations focused almost exclusively on mitigation and adaptation was not prioritized. This changed in the 2000s as developing countries demanded adaptation funding to deal with the immediate and future impacts of climate change. From 2006 onwards the call for adaptation assistance was backed by the fourth IPCC report and a broad and strong transnational "climate justice" movement that included major international development NGOs and international organizations. Policy and academic literature during this period emphasized the overlap between adaptation and development activities, even though climate financing had to be distinct from development assistance for political reasons in the UNFCCC negotiations.

Climate change and migration

This section examines the linkages between climate change and migration. It also encompasses the development of linkages between climate change and displacement and climate change and refugees. There is a significant distinction between refugees, displaced peoples, and migrants. A refugee is someone with "a well-founded fear of being persecuted for reasons of race, religion, nationality, membership of a particular social group or political opinion, is outside the country of his nationality and is unable or owing to such fear, is unwilling to avail himself of the protection of that country."[65] Displaced people are a broader category of persons. They are those forced to move due to any reason—persecution and/or natural disaster related—and may be internally or internationally displaced.[66] Finally, migrant is the broadest category. The UN statistical division defines an international migrant as a person who moves to a country other than that of his or her usual period for at least three months.[67] This definition includes all individuals irrespective of the causes, voluntary or involuntary, and the means, irregular or irregular, used to migrate.[68] These categories mark important differences which are entrenched in migration and refugee law and the institutional separation of IOM and UNHCR. Despite these differences it makes sense to examine the climate change–migration, climate change–displacement, and climate change–refugees issue-linkages collectively as they relate to overlapping academic debates and policy actors and rotate around the central question: Does climate change cause people to move?

An explicit substantive issue-linkage between environmental change and refugees was first widely publicized in the 1970s. Lester Brown of the World Watch Institute argued severe environmental change may force people to move, and thus become "environmental refugees."[69] The term was picked up in other policy literature and entered into common usage in a 1985 UNEP policy paper by Essam El-Hinnawi.[70] However, it was not until 1988 that a World Watch Institute paper claimed that climate change, rather than environmental change, could generate "environmental refugees" and offered an estimate of the numbers.

The first IPCC report in 1990 endorsed a causal link between climate change and displacement. The report predicted that the greatest single consequence of climate change could be "millions of people displaced by shoreline erosion, coastal flooding and severe drought."[71] It argued that "environmental refugees, people displaced by degradation of land, flooding or drought are becoming a much larger factor in many developing countries" and that even a "modest rise in global sea-levels could produce tens of millions of such refugees."[72] The first IPCC report took an alarmist view: climate change would cause millions to move and for this reason governments needed to act urgently.

This claim was supported by the work of Norman Myers, an environmental scientist, who published two cornerstone articles in the 1990s, which influenced the framing of this issue for the next two decades. In an article in *BioScience* he claimed that a "large number of environmental refugees could be among the most significant of all upheavals" caused by climate change.[73] He argued that there were already 10 million environmental refugees and there would be another 150 million in 2050 due to climate change. He later increased this estimate to 250 million.[74] His claims and these figures became the foundation of the "maximalist" perspective and have been repeatedly cited in subsequent academic and policy literature on climate change and displacement.

On the other hand "minimalists" scholars problematized the issue-linkage arguing that there was not a linear, deterministic causal link between climate change and displacement.[75] They critiqued Myers's work for its problematic assumptions, flawed methodology, and incorrect use of the term refugee and negative stereotypes of refugees.[76] They highlighted that numbers of so-called "environmental refugees" were a rough estimate that he arrived at by examining the impact of predicted sea-level rise on populations living in coastal areas and then estimating the projected future population living in those areas (such as the population of coastal Bangladesh), assuming that all affected populations would have to move. All of these figures were based on very rough estimates—of sea-level rise and population growth—and did not take into account the multiple reasons why people might migrate and do so before environmental conditions become dire. Yet even in the most extreme cases not all people move. In fact, it is often the most vulnerable that are left behind—such was the case with Hurricane Katrina. Minimalists argued that migration is a multi-causal, complex phenomenon and you cannot separate out climate change—or environmental change—as a distinct driver of migration except in very extreme cases.[77]

Furthermore, Myers presented refugees in an extremely negative light: they constituted a security threat and brought crime, civil disorder, social upheaval, and violence of many sorts as well as huge financial costs.[78] Others noted he was incorrect to use the term refugee to refer to people displaced by natural disasters and/or climate change.[79] A refugee is defined in international law by the International Convention on Refugees (1951) as someone with:

> A well-founded fear of persecution based for reasons of race, religion, nationality, membership of a particular social group or

> political opinion, is outside his country of nationality and is unable or owing to such fear, is unwilling to avail himself of the protection of that country.[80]

Displacement by natural disaster does not constitute refugee status. This misuse of the refugee category is problematic as refugees are offered special protection (*non-refoulement*) by the international community, which other displaced peoples and migrants are not. The interchanging use of climate "refugee," "displaced person," or "migrant" was common in the academic and policy literature. I refer to all three terms following the terminology of the organization or individual cited.

However, this alarmist argument, regardless of its methodological flaws and problematic assumptions, gained traction in the 2000s. Myers's work became immensely popular and widely cited by other academics and environmental NGOs and lobby groups who sought to persuade governments to take action on climate change.[81] In 2007 climate change displacement began to be a more mainstream policy concern, fueled by the publication of the fourth IPCC report, the Al Gore documentary and the broadening of the climate change regime to include adaptation. Al Gore's documentary showed images of the extraordinarily high neap-tides in Tuvalu, a low-lying Pacific Island state, and claimed mistakenly that New Zealand was accepting "climate refugees" from Tuvalu. Also in 2007 Christian Aid and ActionAid, international development NGOs, published reports on climate change displacement.[82] Christian Aid's report, *The Human Tide: The Real Migration Crisis,* built on Myers's work and predicted that a billion people would be displaced by 2050. These figures were widely cited by academics, policy-makers, and media.

In 2008 and 2009 there was a flurry of NGO reports on "climate refugees." These included: Greenpeace (2008): *Blue Alert, Climate Migrants in South Asia;* Environmental Justice Foundation (2008): *No Place Like Home: Where Next for Environmental Refugees?*, Norwegian Refugee Council (NRC) (2008): *Future Floods of Refugees: A Comment on Climate Change, Conflict and Forced Migration*, NRC (2008): *Climate Changed: A People Displaced*, Environmental Justice Foundation (2008): *No Place Like Home: Where Next for Climate Refugees*, Office for the Coordination of Humanitarian Affairs (OCHA), Internal Displacement Monitoring Centre, NRC (2009): *Monitoring Disaster Displacement in the Context of Climate Change*, and Oxfam (2009): *The Future Is Here: Climate Change in the Pacific.* Claims that millions would be displaced were backed up by Kofi

Annan's *Global Humanitarian Forum* that predicted that 78 million people would be displaced by 2030 in the *Anatomy of a Silent Crisis.*[83]

The UN also played a role in promoting the linkage between climate change and displacement. The United Nations University, Institute of Environmental and Human Security (UNU-EHS), based in Bonn, conducted a prominent study of environmental migration between 2007 and 2009. The "Environmental Change & Forced Migration Scenarios" (EACH-FOR) project was funded by the EU, covered 23 countries and found that climate change was leading to migration. Their work gained widespread media attention and was the basis for later claims by the UN that climate change would create millions of "eco-refugees."[84] The UNU-EHS, as a university and part of the UN, was able to effectively bridge the NGO, UN, and academic sphere and initiate a targeted advocacy campaign for the recognition of the impacts of climate change on migration in the UNFCCC.

In sum, between late 2007 and 2009 a broad transnational "climate justice" movement formed of international development, environmental and human rights NGOs, local civil society groups, and academics. They identified "climate refugees" as the tangible victims of government inaction to mitigate carbon emissions. During the UNFCCC negotiations at Copenhagen an "International Campaign on Climate Refugees' Rights" was launched, and there was a march by a civil society group, "Climate No Borders," which demanded open borders to enable climate refugees free movement. American documentary maker Michael Nash also launched an emotive feature-length documentary in Copenhagen called *Climate Refugees.* The film interviewed what it called "climate refugees" in Bangladesh and Tuvalu as well as experts and politicians including US speaker of the House of Representatives, Nancy Pelosi, Senator John Kerry, and Kenyan Nobel Peace Prize winner, Wangari Maathai. The film predicted that millions would be displaced from Asia and Africa to North America and Europe and thus climate change should be seen as a major security threat. The film launch was supported by CARE International—despite its negative stereotyping of refugees and overly deterministic argument.

Governments adopted this alarmist view of climate change migration as it fitted with their conception of high politics. Leaders securitized climate change as it communicated a "certain gravitas … and warrants a policy response commensurate in effort if not in kind with war."[85] For example, Margaret Beckett, the UK foreign secretary, warned at the UN Security Council in April 2007 that climate change was a security threat that could cause "flooding, disease and famine, and from that migration on an unprecedented scale."[86] The US Quadrennial Defense Review of 2010 identified climate change as a future cause of conflict and mass migration.[87] The most vulnerable countries to climate change also used an alarmist discourse to alert and mobilize others to the urgent need for action. The Bangladeshi finance minister in December 2009 argued that climate change would lead to greater migration and asked all the country's development partners for the naturalized right to move as "managed migration is always much better."[88] However, most low-lying states were clear they did not want to leave and did not see themselves as refugees.[89] As climate change became a more important issue it became securitized by governments and the climate change–migration linkage gained traction.

The media also endorsed the alarmist view. They covered the reports published by UNU-EHS and other international NGOs.[90] They sought visible evidence of climate change and journalists traveled to Bangladesh and low-lying islands, reported to be "sinking," to "find" climate refugees. *The Times*, for example, sent a reporter and a cameraman to the Carteret Islands in Papua New Guinea in the lead-up to Copenhagen to interview locals who they claimed were the first "climate change refugees."[91] Yet the stories of their migration were more complex than presented. It is not clear whether the Carterets were sinking due to climate change or due to tectonic plate subversion.

However, some humanitarian organizations took issue with the negative stereotyping of refugees that the alarmist rhetoric entailed. In 2008 the international humanitarian coordinating committee—the Inter-Agency Standing Committee (IASC)—established a working group on climate change.[92] This group, instigated by Madeleine Helmer, director of the IFRC Climate Change Center, and supported by IOM, wrote a number of submissions on the potential humanitarian impacts of climate change.[93] They endorsed the message that climate change could result in mass displacement but sought to nuance it. In a series of submissions the IASC argued that climate change was not likely to lead to mass displacement from the Global South to the Global North.[94] Instead it was likely to result in internal displacement, which would be covered by the *Guidelines on Internally Displaced Peoples.* Humanitarian agencies were still concerned for the minority who were displaced across international borders by extreme weather events as there was no international legal framework to protect them. Lobbying for such a framework became the focus for several organizations, including UNHCR, NRC, and academics at the UNU-EHS, and will be examined in more depth in subsequent chapters.

In the mid-2000s an additional issue-linkage emerged: migration could also be a voluntary form of adaptation to climate change. A number of migration academics, policy-makers and activists challenged the alarmist rhetoric of climate change migration as a security threat described above.[95] They argued that it was not necessarily problematic that people migrated. In fact, for centuries migration has been an important adaptation and survival strategy to all sorts of environmental pressures. In the lead-up to the Copenhagen UNFCCC meeting the IFRC, the NRC, IOM, and researchers at the UNU-EHS formed an alliance to advocate for migration as an adaptation strategy. They lobbied UNFCCC negotiators actively at Copenhagen (2009) and at Cancún (2010) to include migration as adaptation in the final agreement.[96] UNU-EHS representatives for instance sought meetings with state delegates, held press meetings, and gained media coverage.

At Cancún they succeeded: states included migration as a form of adaptation in the final text. The previously mentioned lobby group had drafted the text of paragraph 14(f), which invited all parties to undertake and enhance adaptation actions including "measures to enhance understanding, coordination and cooperation related to national, regional and international climate change induced displacement, migration and planned relocation, where appropriate."[97] Paragraph 14(f) was lauded as a major achievement within the migration and humanitarian community—although it was not a particularly controversial statement for states. The work of this coalition represented a reframing of migration within the climate change regime as a positive adaptation strategy. This stance was also supported by the UK government's publication in 2011 of the Foresight Report on *Migration and Global Environmental Change, Future Challenges and Opportunities*, and the creation of a new Climate Change and Migration Alliance in the UK which brought together environment, refugee and migration advocates.[98]

In summary, the link between climate change and migration has its origins in the "environmental refugee" policy and academic debate of the 1970s, 1980s, and 1990s. The terms "climate refugee" and "climate migrant" became commonly used terms in the 2000s due to the work of international environmental and development NGOs. By 2007 there was a strong issue-linkage between climate change and migration. The linkage was promoted and documented in the reports and advocacy work of international NGOs such as CARE International, Oxfam, Christian Aid, ActionAid, and Greenpeace. Local advocacy movements (such as the Bangladeshi climate refugees and the UK "Climate No Borders" anarchist group) also endorsed the linkage as it resonated with calls for climate justice. There was high media coverage of the issue and senior politicians—from Nancy Pelosi to Margaret Beckett—endorsed the issue-linkage.

Conclusion

Unlike other examinations of the climate change regime, this chapter focused on how adaptation emerged in the 2000s as a central part of the climate regime. The African COP in Nairobi (2006) pushed adaptation onto the agenda with the Nairobi Work Programme on adaptation; this was followed by the Bali Action Plan (2007), and the Cancun Adaptation Framework (2010). These action plans called for enhanced action on adaptation and argued it should be afforded equal attention in the negotiations. In addition, donor states promised "new and additional" finances for climate adaptation and mitigation in developing countries. At Copenhagen they pledged US$30 billion per year between 2009 and 2013 for a fast track fund; this would increase to US$100 billion per year by 2020. Many of the latter funds were to go through the newly established Green Climate Fund. The inclusion of adaptation opened the door for a number of non-environmental international organizations to engage with climate change.

The chapter then identified how epistemic communities forged new substantive issue-linkages between climate change and migration and climate change and development. It suggested that a linkage between climate change and development has been strong since the UNFCCC's inception as industrialization is widely believed to lead to climate change. However it was only in the mid-2000s that adaptation became linked to development and migration issue-areas. During this time, a network of NGOs, civil society groups, and academics emphasized how climate change would affect people—leading them to move and undermining their human development. The next chapters now examine in depth the role of three organizations in elaborating and acting on these linkages.

Notes

1 Robert Keohane and David Victor, "The Regime Complex for Climate Change," *Perspectives on Politics* 9, no. 1 (2011): 7–23.
2 Paul Harris, *What's Wrong with Climate Politics and How to Fix It* (Cambridge: Polity, 2013).

3 Frank Biermann, Philipp Pattberg, and Fariborz Zelli, *Global Climate Governance Beyond 2012, Architecture, Agency and Adaptation* (Cambridge: Cambridge University Press, 2010).
4 Asa Persson, Richard Klein, Clarisse Siebert, Aaron Atteridge, Benito Muller, Juan Hoffmaister, Mihael Lazarus, and Takeshi Takama, *Adaptation Finance under a Copenhagen Agreed Outcome* (Stockholm: Stockholm Environment Institute, 2009).
5 Richard Klein, "An IPCC Perspective on Climate Adaptation, with Special Reference to Coastal Zones," in *Proceedings of SURVAS Expert Workshop on European Vulnerability and Adaptation to impacts of Accelerated Sea-Level Rise (ASLR)* (Hamburg, Germany: 2000).
6 Lisa Schlipper, "Conceptual History of Adaptation in the UNFCCC Process," *Review of European, Comparative and International Environmental Law* 15, no. 1 (2006): 82–92.
7 The 1992 UNFCCC urged developed countries to assist "developing countries that are particularly vulnerable to the adverse impacts of climate change in meeting costs of adaptation to those adverse effects" (Article 4.4). The UNFCCC agreement also called for cooperation in "preparing for adaptation to the impacts of climate change; develop and elaborate appropriate and integrated plans for coastal zone management, water resources and agriculture, and for the protection and rehabilitation of areas, particularly in Africa, affected by drought and desertification, as well as floods" (Article 4.1). Mark Pelling, *Adaptation to Climate Change, from Resilience to Transformation* (Abingdon: Routledge, 2011).
8 Klein, "An IPCC Perspective on Climate Adaptation, with Special Reference to Coastal Zones," 115.
9 UNFCCC 2014, "Adaptation," http://unfccc.int/adaptation/items/4159.php.
10 Joshua Busby, *Moral Movements and Foreign Policy* (Cambridge: Cambridge University Press, 2010).
11 UNFCCC 2014, "Adaptation," http://unfccc.int/adaptation/items/4159.php.
12 Kati Kulovesi, Maria Gutierrez, Peter Doran, and Mique Munoz, "UN 2006 Climate Change Conference: A Confidence-Building Step?," *Climate Policy* 7 (2007): 255–261.
13 UNFCCC 2007, *The Nairobi Work Programme on Impacts, Vulnerability and Adaptation to Climate Change*, http://unfccc.int/resource/docs/publications/nwp_eng.pdf.
14 "Report of the Conference of the Parties on Its Thirteenth Session, Held in Bali from 3 to 15 December 2007," UNFCCC (FCCC/CP/2007/6/Add.1), 14 March 2008.
15 UNFCCC 2014, "Adaptation," http://unfccc.int/adaptation/items/4159.php.
16 IPCC, *The IPCC Fourth Assessment Report (AR4)* (Cambridge: Cambridge University Press, 2007).
17 Samuel Fankhauser and Guido Schmidt-Traub, "From Adaptation to Climate-Resilient Development: The Costs of Climate-Proofing the Millennium Development Goals in Africa," *Climate and Development* 3, no. 2 (2011): 94–113.
18 Matthew Paterson, *Global Warming and Global Politics* (London: Routledge, 1996).

19 Joyeeta Gupta, "Climate Change and Development Cooperation: Trends and Questions," *Current Opinion in Environmental Sustainability* 1 (2009): 207–213.
20 Nina Hall, "Money or the Mandate? Why International Organizations Engage with the Climate Change Regime," *Global Environmental Politics* 15, no. 2 (2015): 79–96; Jennifer Hadden, *Networks in Contention, the Divisive Politics of Climate Change* (New York: Cambridge University Press, 2015).
21 Hall, "Money or the Mandate? Why International Organizations Engage with the Climate Change Regime"; Mike Hulme, *Why We Disagree About Climate Change, Understanding Controversy, Inaction and Opportunity* (Cambridge, Mass.: Cambridge University Press, 2009).
22 The author was present at the 2009 UNFCCC in Copenhagen and had access to the Bella Center, where negotiations were held.
23 "Report of the Conference of the Parties on Its Fifteenth Session, Held in Copenhagen from 7 to 19 December 2009," UNFCCC (FCCC/CP/2009/11/Add.1), 30 March 2010.
24 Ibid. The US$100 billion in additional funding is supposed to come from a range of sources including the private sector, carbon markets and states. See UN, *Report of the Secretary-General's High-Level Advisory Group on Climate Change Financing* (New York: UN, 2010).
25 "Report of the Conference of the Parties on Its Sixteenth Session, Held in Cancun from 29 November to 10 December 2010," UNFCCC (FCCC/CP/2010/7/Add.1), 15 March 2011.
26 "Report of the Conference of the Parties on Its Seventeenth Session, Held in Durban from 28 November to 11 December," UNFCCC (FCCC/CP/2011/9/Add.1), 15 March 2012.
27 Ibid.
28 "Report of the Conference of the Parties on Its Eighteenth Session, Held in Doha from 26 November to 8 December," UNFCCC (FCCC/CP/2012/8/Add.1), 28 February 2013.
29 Barbara Buchner, Martin Stadelmann, Jane Wilkson, Federico Mazza, Anja Rosenberg, and Dario Abramskiehn, *The Global Landscape of Climate Finance 2014* (Climate Policy Initiative, 2014).
30 Carl Bruch and Jordan Diamond, "The International Architecture for Climate Change Adaptation Assistance," in *Climate Change, Adaptation, and International Development, Making Development Cooperation More Effective*, ed. Ryo Fujikura and Masato Kawanishi (London: Earthscan, 2011).
31 Ibid.
32 Ibid.
33 OECD 2015, "OECD Statistics on External Development Finance Targeting Environmental Objectives Including the Rio Conventions," www.oecd.org/dac/stats/rioconventions.htm.
34 Buchner et al., *The Global Landscape of Climate Finance 2014*.
35 Axel Michaelowa and Katharina Michaelowa, "Coding Error or Statistical Embellishment? The Political Economy of Reporting Climate Aid," *World Development* 39, no. 11 (2011): 2010–2020.
36 Bruch and Diamond, "The International Architecture for Climate Change Adaptation Assistance."

37 Jessica Brown, Neil Bird, and Liane Schalatek, "Climate Finance Additionality: Emerging Definitions and Their Implications," in *Climate Finance Additionality: Emerging Definitions and Their Implications* (London: Overseas Development Institute, 2010), 10.
38 Interview with senior policy advisor, Chief Executives Board, UN Secretariat, 15 October 2010, New York.
39 Ibid.
40 Sandeep Sengupta, "International Climate Negotiations and India's Role," in *Handbook of Climate Change and India: Development, Politics and Governance*, ed. Navroz K. Dubash (London and New York: Routledge and Oxford University Press, 2011)
41 Paterson, *Global Warming and Global Politics.*
42 I will not explore their arguments here as they have been dealt with and given ample examination elsewhere. Furthermore, the exact relationship between industrialization, increasing carbon emissions, and global warming is an ongoing matter of scientific debate and study. However, what many mainstream commentators ignore is that science is never "settled." The process of scientific inquiry involves testing and refining all "facts." Even Einstein's theory of relativity is being tested and reworked. There is a strong substantive issue-linkage between development and climate change. For more on the nature of scientific inquiry and the politicization of science in the climate change debate see Hulme, *Why We Disagree About Climate Change, Understanding Controversy, Inaction and Opportunity.*
43 Ibid.
44 Ibid.
45 Richard T.J. Klein, "Climate and Development in Times of Crisis," *Climate and Development* 1, no. 1 (2009): 3–4.
46 UN General Assembly. 96th Plenary meeting, *Report of the World Commission on Environment and Development*, (A/RES/42/187), 11 December 1987.
47 Ibid.
48 Rio Declaration on Environment and Development, UN General Assembly (A/CONF.151/26 [Vol. I], Article 3).
49 Zoe Young, *A New Green Order? The World Bank and the Politics of Global Environment Facility* (London: Pluto, 2002).
50 Ibid.
51 John Magrath, "How Climate Change Made the Development Agenda," IIED, 2014. www.iied.org/how-climate-change-made-development-agenda?utm_source=feedburner&utm_medium=email&utm_campaign=Feed%3A+IIEDblogs+%28International+Institute+for+Environment+and+Development+blogs%29.
52 The organization closed in 2010 due to a lack of funding. See Global Humanitarian Forum 2015, "Global Humanitarian Forum," www.ghf-ge.org/.
53 Frances J. Moore, "'Doing Adaptation': The Construction of Adaptative Capacity and its Function in the International Climate Negotiations," *St Antony's International Review* 5, no. 2 (2010): 66–88.
54 IPCC, *Climate Change, the IPCC Scientific Assessment* (Cambridge: Cambridge University Press, 1990b).

55 IPCC, *The IPCC Fourth Assessment Report (AR4)*.

56 There are a number of definitions of adaptation in the literature. Siri Eriksen et al., "When Not Every Response to Climate Change Is a Good One: Identifying Principles for Sustainable Development," *Climate and Development* 3 (2011): 7–20.

57 Interview with senior OCHA representative working on climate change, 19 March 2010, Geneva.

58 UNDP 2015, "Adapting to Climate Change," www.undp.org/content/undp/en/home/ourwork/environmentandenergy/strategic_themes/climate_change/focus_areas/adapting_to_climatechange/.

59 WHO 2015, "Health Adaptation to Climate Change," www.who.int/globalchange/climate/gefproject/en/. For another example the World Bank sees adaptation as making "the effects of climate change less disruptive and spare the poor and the vulnerable from shouldering an unduly high burden." World Bank 2011, "Economics of Adaptation to Climate Change," www.worldbank.org/en/news/feature/2011/06/06/economics-adaptation-climate-change.

60 Persson et al., *Adaptation Finance under a Copenhagen Agreed Outcome.*

61 Siri Eriksen and Jeremy Lind, "Adaptation as a Political Process: Adjusting to Drought and Conflict in Kenya's Drylands," *Environmental Management* 43 (2009): 817–835.

62 Klein, "Climate and Development in Times of Crisis."

63 Moore, "'Doing Adaptation': The Construction of Adaptative Capacity and its Function in the International Climate Negotiations."

64 Persson et al., *Adaptation Finance under a Copenhagen Agreed Outcome.*

65 International Convention relating to the Status of Refugees, Art. 1A(2), 1951 as modified by the 1967 Protocol. Note that states have established and use a number of other definitions to determine refugee status. The Organization of African Unity (OAU) adds that a refugee is also any person compelled to leave his or her country "owing to external aggression, occupation, foreign domination or events seriously disturbing public order in either part or the whole of his country of origin or nationality." *Convention Governing the Specific Aspects of Refugee Problems in Africa*, Article 1 (2), Addis Ababa, 10 September 1969. The 1984 Cartagena Declaration also includes persons who flee their country "because their lives, security or freedom have been threatened by generalized violence, foreign aggression, internal conflicts, massive violations of human rights or other circumstances which have seriously disturbed public order." Cartagena Declaration on Refugees, 22 November 1984, Cartagena de Indias (3).

66 The Guiding Principles on Internal Displacement defines IDPs as "Persons or groups of persons who have been forced or obliged to flee or to leave their homes or places of habitual residence, in particular as a result of or in order to avoid the effects of armed conflict, situations of generalized violence, violations of human rights or natural or human-made disasters, and who have not crossed an internationally recognized State border." "Guiding Principles on Internal Displacement," UN (UN Doc E/CN.4/1998/53/Add.2), 11 February 1998.

67 UN Statistical Division, "International Migration—Concepts and Definitions." http://unstats.un.org/unsd/demographic/sconcerns/migration/migrmethods.htm
68 Note that the UN and IOM tend to adopt a narrower definition of migrants, to focus on "long-term" international migrants. This is "a person who moves to a country other than that of his or her usual residence for a period of at least a year." Ibid.
69 James Morrissey, *Environmental Change and Forced Migration* (Oxford: Refugee Studies Centre, 2009).
70 Ibid.
71 IPCC, *Climate Change the IPCC Impacts Assessment Report* (Canberra: IPCC, 1990a).
72 Ibid.
73 Norman Myers, "Environmental Refugees in a Globally Warmed World," *BioScience* 43, no. 11 (1993): 752–761.
74 Norman Myers, "Environmental Refugees," *Population and Environment: A Journal of Interdisciplinary Studies* 19, no. 2 (1997): 167–182.
75 Richard Black, "Environmental Refugees: Myth or Reality?," *New Issues in Refugee Research* Working Paper No. 34 (2001): 1–19; Astrid Suhrke, "Environmental Degradation and Population Flows," *Journal of International Affairs* 47, no. 2 (1994): 473–496.
76 Black, "Environmental Refugees: Myth or Reality?"; Jane McAdam, "Environmental Migration Governance," in *Global Migration Governance*, ed. Alexander Betts (Oxford: Oxford University Press, 2009).
77 Migrants move because of a range of interconnected social, political, and economic factors. In addition, some argued that it was problematic to analytically separate the environment from society as a causal factor of migration as "perceptions of nature are socially embedded." A key claim was that the relationship between the environment—let alone climate change—and human mobility is "inherently ambiguous." Alexander Betts, "Substantive Issue-Linkage and the Politics of Migration," in *Arguing Global Governance*, ed. Corenliu Bjola and Markus Kornprobst (Abingdon: Routledge, 2010), 7.
78 Myers, "Environmental Refugees in a Globally Warmed World."
79 Jane McAdam, "Refusing 'Refuge' in the Pacific: (De)Constructing Climate Induced Displacement in International Law," in *Migration, Environment and Climate Change*, ed. Étienne Piguet, Antoine Pécoud, and Paul De Guchteneire (Paris: UNESCO, 2010)
80 1951 Convention Relating to the Status of Refugees.
81 Friends of the Earth, *Gathering Storm: The Human Cost of Climate Change* (Amsterdam: Friends of the Earth, 2000); Friends of the Earth, *Our Environment, Our Rights: Standing up for People and the Planet* (Amsterdam: Friends of the Earth, 2004).
82 ActionAid 2012, *Unjust Waters: Climate Change, Flooding and the Protection of Poor Urban Communities: Experiences from Six African Cities*, www.actionaid.org.uk/sites/default/files/doc_lib/unjust_waters.pdf.
83 Global Humanitarian Forum 2009, *The Anatomy of a Silent Crisis*, www.ghf-ge.org/human-impact-report.pdf.
84 Stefan Lovgren, "Climate Change Creating Millions of 'Eco Refugees' UN Warns," *National Geographic News*, 18 November 2005.

85 John Barnett, "Security and Climate Change," *Global Environmental Change* 13 (2003): 7–17.

86 Margaret Beckett, "The Case for Climate Security," *The RUSI Journal*, 152, no. 3 (2007): 54–59. The Stern Review forecasted that "greater resource scarcity, desertification, risks of drought and floods, and rising sea levels could drive millions of people to migrate." Nicholas Stern, *The Stern Review on the Economics of Climate Change*, 2006, http://webarchive.nationalarchives.gov.uk/+/http:/www.hm-treasury.gov.uk/independent_reviews/stern_review_economics_climate_change/stern_review_report.cfm.

87 US Department of Defense, *Quadrennial Defense Review*, (Department of Defense: Washington, 2010). There was a side event at Copenhagen entitled "Climate Change and International Security" at which the North Atlantic Treaty Organization (NATO) secretary general spoke alongside the African Union commission chairperson and the EU presidency, Swedish minister for foreign affairs. One of the main themes of this presentation was the potential security risk of climate change refugees. NATO also took an alarmist, securitized view of climate change migration, which could trigger the need for humanitarian intervention and lead to regional instability

88 "Bangladesh Climate Aid: 'We Are Getting Peanuts,'" *The Guardian*, 2009. www.theguardian.com/environment/video/2009/dec/02/bangladesh-climate-aid.

89 McAdam,"Refusing 'Refuge' in the Pacific: (De)Constructing Climate Induced Displacement in International Law."

90 Tom Zeller Jr., "Making the Case for Climate as a Migration Driver," *New York Times*, June 14, 2009. www.nytimes.com/2009/06/15/business/energy-environment/15iht-green15.html?pagewanted=all&_r=1.

91 "The Carteret Islands," *The Telegraph*, 2009. www.telegraph.co.uk/news/earth/carteret-islands/.

92 Nina Hall, "A Catalyst for Cooperation? The Inter-Agency Standing Committee and the Humanitarian Response to Climate Change," *Global Governance*, forthcoming (2016).

93 Interview with OCHA staff member, 20 March 2010, Geneva. Interview with ISDR staff member and IASC member, 13 January 2012, Wellington, New Zealand.

94 The Informal Group on Migration/Displacement and Climate Change of the IASC, "Climate Change, Migration and Displacement: Who Will Be Affected?" *Working Paper*, 31 October 2008. IASC Working Group on Climate Change, *Comments and Proposed Revisions to the negotiating text prepared by the Chair of the UNFCCC Ad Hoc Working Group on Long Term Cooperative Action,* June 2009.

95 Lorraine Elliot, "Climate Migration and Climate Migrants: What Threat, Whose Security?," in *Climate Change and Displacement: Multidisciplinary Perspectives*, ed. Jane McAdam (Oxford: Hart Publishing, 2010)

96 Koko Warner, *Climate Change Induced Displacement: Adaptation Policy in the Context of the UNFCCC Climate Negotiations* (Geneva, Switzerland: UNHCR—Division of International Protection, 2011).

97 "Report of the Conference of the Parties on Its Sixteenth Session, Held in Cancun from 29 November to 10 December 2010," UNFCCC (FCCC/CP/2010/7/Add.1), 15 March 2011.

98 Government Office for Science and Foresight, *Foresight Report Migration and Global Environmental Change, Future Challenges and Opportunities* (London: Government Office for Science and Foresight, 2011). Discussions with Hannah Smith, Climate Change and Migration Alliance Coordinator, COIN, May 2010 and February 2011.

2 UNHCR and climate change

- **UNHCR's evolving mandate**
- **UNHCR and climate change (2000–2015)**
- **Conclusion**

UNHCR occupies a unique place in global governance: it is the international provider of protection for refugees and stateless people.[1] The office was created in 1950 with a narrow legal protection mandate and studies have emphasized how it has autonomously evolved to become one of the world's primary providers of humanitarian services.[2] In recent years many NGOs and academics have argued that climate change will force people to move, from low-lying coastal lands, drought-prone areas, and other at risk areas, as discussed in the previous chapter. Some have also argued for a new legal convention to protect those displaced.[3] Yet there is no academic literature exploring if and how UNHCR has expanded beyond the established refugee regime to respond to climate displacement, which this chapter addresses.

This chapter begins by outlining UNHCR's original mandate and its subsequent expansion. It then focuses on how UNHCR dealt with environmental issues in the 1990s. The third and most substantive section illustrates how UNHCR engaged with climate change between 2000 and 2015. It focuses on changes in rhetoric, policy, and structure at headquarters, and at the operational level in Kenya. Finally it examines mandate expansion and demonstrates how states resisted the High Commissioner's efforts to expand into protection of people who are displaced by climate change and/or affected by natural disasters. Overall, it suggests that UNHCR's response to climate change was shaped by staff interpretation of their core refugee protection mandate.

UNHCR's evolving mandate

Original mandate

UNHCR was created in 1950 with a mandate to protect thousands of people displaced by war and those who fled Eastern Europe.[4] In 1951, a year after the establishment of UNHCR, states signed the Convention Relating to the Status of Refugees (Refugee Convention), which set out the definition and rights of refugees and gave UNHCR supervisory responsibility for the convention. Article 35 of the convention stipulated that: "the contracting states undertake to cooperate with the Office of the UNHCR … in the exercise of its functions and shall in particular facilitate its duty of supervising the application of the provisions of this Convention."[5] UNHCR was also given the authority and power to "promote the conclusion and ratification of international Conventions" to protect refugees.[6] UNHCR was established with a legal mandate to supervise the application and development of international refugee law.[7]

Refugees were defined narrowly in the convention, reflecting the postwar context. A refugee is someone with "A well-founded fear of being persecuted for reasons of race, religion, nationality, membership of a particular social group or political opinion, is outside the country of his nationality and is unable or owing to such fear, is unwilling to avail himself of the protection of that country."[8] For UNHCR and as a matter of law, a refugee is distinguished from an economic migrant or someone displaced by a climate change related disaster (flooding, drought, sea-level rise) or any other natural disaster (earthquake, tsunami) as they face persecution. Refugee status is a specific, legal category of persons and not to be confused with the broader conception of a refugee as anyone fleeing his/her country and seeking refuge, used often by media, NGOs, and some academics.[9] This legal categorization of refugees is at the heart of UNHCR's mandate and identity.

The 1951 Refugee Convention took on an "almost constitutional character" within UNHCR.[10] The office's supervisory status gave the High Commissioner "considerable moral authority and legitimacy" even though this position had little political legitimacy.[11] In particular the agency was mandated with a unique and central role in supervising the implementation of international refugee law and offering protection to refugees.[12] UNHCR has legal authority to judge state behavior and challenge states whose policies undermine refugee law and threaten refugees' rights.[13] States on the other hand were expected to cooperate with the Office of the High Commissioner in its activities.[14] These legal functions gave the agency a "unique identity and considerable independence," according to UNHCR's director of international protection.[15] UNHCR's legal and moral authority is widely recognized in scholarship and is often closely correlated with its relative autonomy.[16]

The Division of International Protection upholds the core identity of UNHCR. Its role is to interpret the Refugee Convention and protect the refugee status. It is the oldest division in UNHCR and hosts the "traditionalists," who tend to "view the UNHCR and itself as the refugee's lawyer and as the protector of refugee rights under international law."[17] UNHCR inherited the Legal Protection Division from its immediate predecessor the International Refugee Office, which had developed the notion of individualized refugee protection.[18] In the first decades of UNHCR the Legal Protection Division (based at the headquarters in Geneva) was the dominant focus of the agency's work. The agency had no operational staff working in the field to offer assistance or protection in humanitarian crises. The Legal Protection Division is responsible for determining the legal criteria of refugee status and ensuring that states uphold their obligations.

UNHCR's organizational culture reflects the agency's legal standing as the protector of refugees. Staff in UNHCR are wedded to the mandate and strongly committed to refugee protection. UNHCR staff "almost universally believe in the principles of the mandate."[19] Scholars have argued that "there exists no other UN agency where values and principled ideas are so central to the mandate and *raison d'être* of the institution, or where some committed staff members are willing to place their lives in danger to defend the proposition that persecuted individuals need protection."[20] An independent review of UNHCR's organizational culture in 2005 found that staff had a "very strong collective identity" and almost universally believed in the principles of the organization's core mandate.[21] States also acknowledge that UNHCR has a "normative setting" as UNHCR staff are "the guardians of the Refugee Convention" and "have the mandate to help these people, therefore you [states] can help us achieve this goal."[22] UNHCR has one of the narrowest original mandates of any UN organization and staff are deeply loyal to this, which makes it a more conservative organization when faced with new issue-areas.

Mandate expansion

UNHCR has used this moral legitimacy to expand significantly and autonomously from its original mandate. The agency started life boxed in with a limited mandate of protection for European refugees who were displaced pre-1951.[23] It had restricted funding, little support from the United States, a small staff of 34 and no operational activities. The United States was frustrated that its candidate did not win the election to be the first High Commissioner and thus refused to fund the agency in the first five years of its life.[24] UNHCR staff body was so small that they were all able to gather around a single piano and sing carols at the office's Christmas party in 1951. Under the 1950 statute the office has a mandate to act "under the authority of the General Assembly," which it had to renew every five years until 2003.[25] In addition, it has to report annually to the General Assembly through the Economic and Social Council (ECOSOC) and follow policy directives issued by either of these two bodies. In the late 1950s and early 1960s the agency began to expand its mandate beyond Europe and into Asia (Hong Kong) and Africa (Algeria) through the use of "good-offices," which were established to deal with specific refugee flows. Initially these UNHCR operations outside of Europe were ad-hoc expansions, which had to be endorsed by member states. However, in 1967 states agreed to an additional protocol to the UNHCR convention, which removed the temporal and geographical limitations of the original refugee definition.[26] UNHCR was also given supervisory status over the 1961 Convention Relating to the Status of Stateless Persons (Statelessness Convention) and thus increased the scope of its moral legitimacy.[27]

In the 1970s, 1980s, and 1990s UNHCR continued to expand geographically and in scope. In the 1970s, UNHCR operations moved into Southeast Asia, Latin America and later the Middle East and now the agency operates globally. Correspondingly, there was a rapid expansion of UNHCR's budget and staff: it grew from a budget of several million and a few hundred staff in 1971, to thousands of staff and a budget of US$500 million in 1980. In 2013 UNHCR had an annual budget of almost US$4 billion, along with US$1.3 billion for supplementary appeals, and at the beginning of 2015 there were over 9,300 staff.[28] UNHCR increased the range of its activities to include prevention, early warning and development assistance, and in-country protection.[29] In the 1990s and 2000s it also expanded its definition of "persons of concern" and now offers protection and/or assistance to returnees, asylum-seekers, internally displaced persons (IDPs), and other persons threatened with displacement. This was a significant development as it meant UNHCR had a much broader range of people it catered to and activities it provided. The office's mandate for these groups of people evolved from "residual powers" vested in the UNHCR statute, which enable it to engage "in such additional activities as the General Assembly may determine."[30]

Interestingly, UNHCR's expansion into IDPs was driven by two of UNHCR's top donors, the United States and the Netherlands, not by staff.[31] Many staff, academics, and some NGOs asserted that expansion into assisting IDPs could undermine UNHCR's ability to deliver on refugee protection.[32] Expansion has often met resistance in UNCHR because of staff's primary concern to uphold the Refugee Convention. Moreover, UNHCR's mandate for IDPs, returnees, and other persons "threatened with displacement" is unlike its mandate for refugees and stateless persons as it is not based on an international convention but on state consent. UNHCR's operations with these new groups are seen as additional add-ons or extra-mandates that must not detract from or dilute the agency's core mandate for refugees.[33] In sum, UNHCR's protection role for refugees was of a higher importance and priority because it was stipulated in the statute and by the convention, and not a General Assembly resolution which could be changed from one year to the next.

Alongside this expansion in scope, the agency became a more operational organization driven by emergencies, as well as refugee flows. These changing priorities were reflected in structural changes. In the 1980s and 1990s the Division of Operations grew significantly and rivaled the Division of International Protection. The director of the Division of Operations was promoted to the same level as the director of the Division of International Protection, as demonstrated in the organigram in Figure 2.1.[34] This placed a higher value on operational activities vis-à-vis legal protection and led to tensions within UNHCR between these two goals.[35] While the Division of Protection aims to defend the rights of refugees, often against states, the Division of Operations is more liable to accede to state pressures as its goal is to deliver humanitarian assistance. A refugee advocate argued that "international law, which provides the essential framework of rule and principle for the protection of refugees, has been relegated to an inferior position vis-à-vis the political concerns of UN Member States."[36] The refugee advocate also stated that the "authority and influence of the Division [of International Protection] has eroded to the point that its advice is now generally sought only after decisions have been made by policy and program staff, and then only to find retrospective legal justification for UNHCR action, which has a number of detrimental consequences."[37] UNHCR hired new staff for the Division of Operations who were not loyal to, or schooled in, UNHCR's refugee law mandate.

UNHCR's activities are closely monitored by states through annual reporting and informal discussions to ensure mandate creep does not

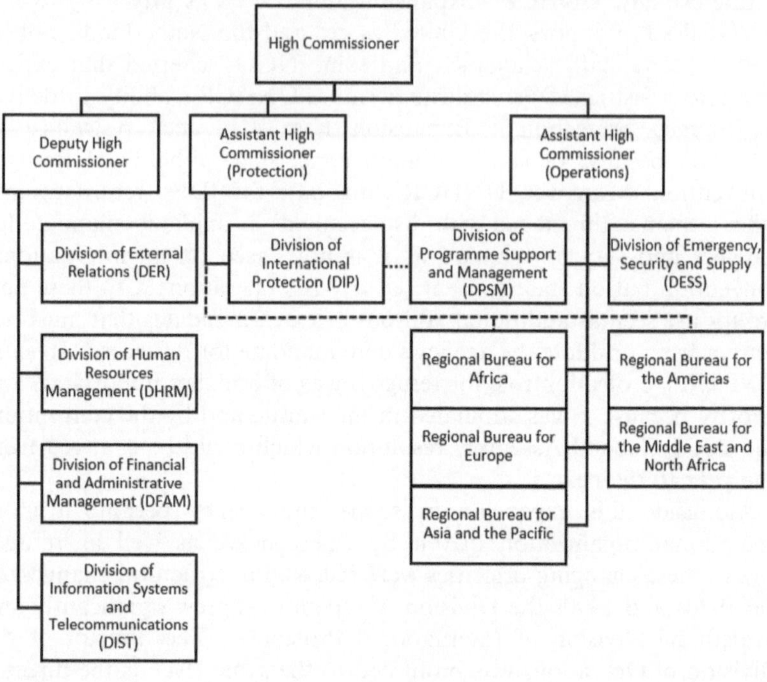

Figure 2.1 UNHCR organigram (2015)
Adapted from: *UNHCR Headquarters Organizational Structure*, www.unhcr.org/4bffd0dc9.html.

occur. While the agency's original mandate came from the General Assembly, states oversee the agency today primarily through the Executive Committee of the Programme of the UNHCR (ExCom) which meets on a regular basis.[38] UNHCR is particularly reliant on its donors to continue its operations and has no permanent funding, which means that long-term, multi-year planning is difficult. Approximately 98 percent of the office's funding is from voluntary contributions.[39] In fact, the agency receives approximately three quarters of its budget from its top ten donors who "exercise significant influence over the work of UNHCR."[40] Meanwhile, many treat the ExCom cautiously and reject anything "that is too specific as they may be bound by it" and do not want to set "prescriptive recommendations for the future" which they cannot meet.[41] As another member state explained: "What they [UNHCR] say and do sets a precedent and leads the debate."[42]

This chapter will now examine how UNHCR's core identity as a protection agency for refugees has shaped its engagement firstly with the issue of environment and then climate change.

UNHCR and the environment (1990–1999)

UNHCR was not created with a mandate to protect those displaced by environmental disaster or climate change. However, in the 1990s, following the end of the Cold War, the agency began to engage with environmental issues. It did this in two ways: 1) the issue of environmental displacement was discussed as a protection issue, and 2) the agency developed an environmental division and environmental impact guidelines. These changes occurred in parallel to the expansion of its humanitarian operations under the leadership of a new High Commissioner, Sadako Ogata (1990–2000). Ogata perceived a need to restore the relevance and raise the profile of UNHCR and therefore made appearances at the UN Security Council and NATO and gained widespread international press coverage for the agency.[43] She also sent UNHCR into Yugoslavia to provide assistance to thousands, and not just those displaced. This section will focus on UNHCR's engagement with environmental issues during this period of operational expansion.

Between 1990 and 1991 UNHCR and the ExCom discussed the issue of international protection for environmentally displaced peoples. This discussion was through a working group, and its report, *Solutions and Protection.* The working group was convened by the High Commissioner, at the request of the ExCom in 1989 and comprised of states, international organizations, and UNHCR staff who were instructed to look at protection and preventive solutions to displacement. They discussed protection for groups beyond UNHCR's core refugee and stateless persons mandate. The working group identified several categories of displacement including: "people forced to leave or prevented from returning because of man-made disasters; persons forced to leave or prevented from returning because of natural or ecological disasters or extreme poverty."[44] It noted that natural or ecological disaster is a cause of migration and "up-rootedness" and could contribute to a flow of persons across borders, although most cases of displacement are likely to be internal. It also stated that "persons fleeing natural or ecological disaster normally have a need for relief assistance rather than protection."[45] However, it was agreed that "the competence of UNHCR does not normally extend to those persons displaced inside their own countries as a result of natural or ecological disaster or extreme poverty."[46] Other international organizations "with appropriate mandates" should respond to the needs of those displaced externally or internally by environmental causes.[47] In sum, UNHCR had no mandate to offer assistance or protection to those fleeing natural or ecological disasters and in the 1990s—despite the trends towards broader operational expansionism—it did not offer humanitarian relief to them.

Meanwhile, in the 1990s UNHCR began to include environmental impact projects within its humanitarian operations. These efforts were part of a growing acknowledgment that refugees had an impact on their host environment, and as Ogata stated, the "relationship between the environment and refugees had long been overlooked."[48] In 1992 UNHCR attended a conference on Migration and Environment in the lead-up to the UN Conference on Environment and Development in Rio de Janeiro. A UNHCR staff member from the Programme and Technical Support Service explained that UNHCR did not traditionally consider environmental impacts in measuring the effectiveness of its programs but that there was a "growing recognition of the need to take into account the use of environmental resources in planning assistance programs."[49] Subsequently, UNHCR established the Engineering and Environmental Services Section within the Division of Operations, hired its first environmental coordinator in 1993, and in 1998 this section published UNHCR's first environmental impact guidelines.[50] In addition, UNHCR developed environmental projects including piloting solar cookers; refugee environmental education; reducing energy use and deforestation in camps; and sourcing environmentally friendlier materials.[51] By the late 1990s the environmental impact of refugees was considered an issue that UNHCR had to deal with routinely in its operational activities. These activities are not the focus of this research as they were framed not as a response to the issue of climate change but to the environment.[52] Moreover, they concentrate on the impact of UNHCR on the environment and not the impact of climate change on individuals of potential concern to UNHCR and its response to those individuals.

In sum, during the 1990s UNHCR did not offer protection to those displaced by any environmental change or disasters. While there was one working group that discussed this issue it stated clearly that protection for this category of people was beyond the mandate of UNHCR. The environmental work that UNHCR engaged in focused on mitigating the environmental impacts of refugee camps, which is not the focus of this book.

UNHCR and climate change (2000–2015)

No engagement (2000–2007)

UNHCR was a late starter on climate change.[53] It did not engage with policy debates on the issue of climate change and displacement until 2007. Meanwhile, the research and policy work on the issue was led by other agencies, such as the NRC, the IFRC, and IOM.[54] During the early and mid-2000s there was increasing media and NGO attention to the plight of so-called "climate refugees."[55] UNHCR did not engage with climate change during this period as it was not deemed an issue of high political salience. The High Commissioner, Ruud Lubbers (2001–2005), was "reacting to the most important issues of the time" and climate change was not one of these during his leadership of UNHCR.[56]

In addition, senior staff did not see climate change as an important priority and were reluctant to engage with the issue, which was perceived as outside their mandate. A UNHCR staff member explained that "few people had any clue within the organization about climate change or whether it was of any interest to them" and there was a degree of "skepticism" about following global trends.[57] UNHCR's main concern was to refute that "climate refugees" were in fact refugees. They argued that according to the 1951 convention refugees were only those fleeing a well-founded fear of persecution based on race, religion, nationality, membership of a particular social group, or political opinion. This was not a controversial stance and the agency was supported by many refugee law experts.[58]

"Climate refugees" posed a threat to UNHCR as they blurred the boundaries between convention refugees and popular conceptions of refugees. UNHCR staff were concerned that engaging in debates over "climate refugees" would be a distraction from protecting the unique legal status of convention refugees. In fact, those in the Division of International Protection—predominantly refugee lawyers—were particularly reticent to engage with climate change and displacement debates as they saw it as a diversion from UNHCR's mandate.[59] In 2005, a new High Commissioner, António Guterres, the former prime minister of Portugal, took office. He was driven by strong political ambitions and from the outset sought to affirm UNHCR's core mandate and also expand the agency's mandate.[60] However, in the first two years of his leadership Guterres did not speak on climate change. UNHCR in the early 2000s and mid-2000s did not respond to the issue of climate change as it was not deemed politically important or sufficiently relevant to its mandate.

Rhetorical change

UNHCR first engaged with the issue of climate change and displacement in 2007. At the annual meeting of ExCom Guterres began his speech by highlighting that the drivers of displacement were changing. He claimed that:

> Almost every model of the long-term effects of climate change predicts a continued expansion of desertification, to the point of destroying livelihood prospects in many parts of the globe. For each centimeter that the sea level will rise, there will be one million more displaced. The international community seems no more adept at dealing with those new causes than it is at preventing conflict and persecution. It is therefore important to examine the reasons, the scale and the trends of present-day forced displacement.[61]

In this speech for the first time he acknowledged a substantive issue-linkage between climate change and displacement. He made this speech because he had a "strong belief that climate change was a new phenomenon that must be dealt with."[62]

The *Official Summary Record of the 2007 Executive Meeting* notes a mixture of responses from states to the High Commissioner's statements. One state was actively supportive (Norway), some states were vaguely supportive and one was completely opposed (Austria). Yet most of the UNHCR staff interviewed claimed that states were completely opposed to UNHCR expansion in this area. As Jeff Crisp, the head of UNHCR's Policy Development and Evaluation Service, stated many key UNHCR donors have "expressed persistent wariness with regard to the organization's expansion, often expressing the opinion that the organization should return to its 'core mandate' which they consider to be that of providing refugees with protection in developing regions."[63] While a select few member states went on the official record endorsing UNHCR's engagement with climate change, the institutional memory and the record of subsequent meetings suggest that the majority of the executive committee did not actively encourage the agency to engage with climate change and displacement. Many states saw ExCom conclusions as setting a precedent that would subsequently bind them.[64]

Nevertheless, over the course of 2008 and 2009 the High Commissioner continued to highlight in high-level speeches, interviews, and articles that climate change would trigger mass displacement.[65] In numerous speeches he emphasized that climate change was one of five new "mega-trends" changing the nature of displacement.[66] He also made climate change a core theme of many speeches and interviews. For instance, in an interview with *The Guardian* he claimed that "Climate change is today one of the main drivers of forced migration, both directly through its impact on the environment—not allowing people to live any more in the areas where they were traditionally living—and as a trigger of extreme poverty and conflict."[67] The agency had begun to shift its position and acknowledge that climate change may lead to new forms of displacement.

Although there was significant rhetorical engagement from the High Commissioner on the issue, there was a lag in policy outputs by UNHCR. It was relatively easy for the High Commissioner to issue statements and make speeches on the changing nature of forced displacement without committing the agency to any particular role in addressing these new flows. However, there was still widespread resistance within UNHCR bureaucracy to the term "climate refugee." Guterres avoided this term and was aware that adopting this language could weaken UNHCR's core mandate. In fact, Guterres stated that "UNHCR has refused to embrace the new terminology of 'climate refugees' or 'environmental refugees,' fearing that this will complicate and confuse the organization's efforts to protect the victims of persecution and armed conflict."[68] Guterres's concern about expansion undermining the core mandate illustrates how staff internalized the office's mandate, and this shaped their response.

The High Commissioner also faced continual state resistance to any expansion of the mandate. When the issue of climate change and displacement was raised by the High Commissioner at the September 2009 ExCom meeting there was a public backlash from some states. The Bangladeshi representative stated that he had "reservations about any enlargement of the Office's responsibility to cover climate change–induced displacement scenarios. UNHCR should remain focused on its mandated areas where it had comparative advantages."[69] Bangladesh was concerned that an expansion of UNHCR activities into climate displacement would give it grounds to operate within Bangladesh, and thus undermine its sovereignty. While Bangladesh now appears to support a revision of UNHCR's mandate an interview with another member state delegate confirmed that many states were opposed to UNHCR expanding into new areas such as climate change induced displacement.[70]

Despite the lack of support from states and many UNHCR staff, the High Commissioner continued to speak on climate change. In December 2009 he made his first appearance at the UNFCCC negotiations in

Copenhagen. He spoke at a number of high-level side events and press conferences on the humanitarian impacts of climate change. At an IASC-convened side event he claimed that "climate change is expected to unseat conflict as the main driver of mass migration in coming years" and had blurred the boundaries between migration and displacement. At Copenhagen he implied for the first time that there was a need for a new framework to protect climate change displaced peoples. He stated that "the international community must develop new mechanisms for the protection of climate refugees."[71] These were bold statements and constituted a shift in UNHCR's position.

The High Commissioner, by suggesting that there was a need for new protection frameworks, was implicitly positioning his agency to provide this protection. These statements were potentially dangerous for UNHCR's core work in refugee protection. They could be interpreted as undermining the legal refugee definition: it was no longer relevant and should be significantly reworked. Some UNHCR staff explained that the High Commissioner made a strategic decision, as an astute politician, to take this proactive stance on climate change, natural disasters, and displacement.[72] He saw that the issue had increased international importance and UNHCR needed to engage with it in order to retain relevance and credibility.[73] This suggests that the High Commissioner was more inclined towards expansionism than the rest of his agency.

In summary, between 2000 and 2007 UNHCR senior leadership did not respond in rhetoric to the issue of climate change. The issue was not deemed to be relevant to UNHCR's core mandate. High Commissioner Guterres first noted the links between climate change and displacement in 2007 and from 2008 onwards the High Commissioner made climate change a regular theme of speeches. Guterres would commonly refer in his speeches to climate change as a new "megatrend" which would lead to new forms of displacement and statelessness, outside the scope of the Refugee Convention, and he called for new protection mechanisms. We will see the development of this rhetoric in subsequent sections of this chapter as Guterres lobbied states for mandate change at ministerial meetings, the Nansen Conference and in the UNHCR Standing Committee.

Structural change

Broader UNHCR involvement with climate change, beyond the High Commissioner, began in 2008. In May 2008 the UN Secretary General called for all UN agencies to establish climate change focal points to prepare for the UNFCCC summit in Copenhagen and to give higher organizational priority to the issue.[74] The High Commissioner established a task force on climate change in 2008 backed by his "desire for the Office to engage fully and effectively in the international discussion on these issues."[75] The brief of this task force included liaising with the executive office to ensure that a consistent UNHCR position on climate change and related issues would be articulated and adjusted as needed, as well as tracking developments relating to climate change as they concern the mandate of UNHCR and providing inputs into the IASC task force on climate change. This task force included a number of people across the agency—from the Department of Operations to International Protection. Essentially, the task force assigned responsibility to staff to work on aspects of climate change within their other routine responsibilities.

However, the task force neither triggered nor constituted organizational change within UNHCR. In fact, there is evidence to suggest there was almost no change at all. Only a few new positions were created to work on climate change and all were temporary and/or part-time. Moreover, some UNHCR staff reported that even though they were supposedly responsible for working on climate change they had neither the time nor the space to work on it.[76] One staff member, for instance, claimed that he "was not encouraged to make substantive changes or to take the issue seriously" as he never had the means or authority to turn it into a serious policy issue.[77] Senior managers did not prioritize climate change and were reluctant to delegate staff to work on it as they did not see the issue as a core concern or priority. One senior staff member even reported that they were "jaded" with the debate.[78]

UNHCR's core climate change work was done by a focal point in the Policy Development and Evaluation Service.[79] However the office holder was given no formal briefing note on this work and the position was passed around over three years to different people in different departments, much like a hot potato. In 2009 UNHCR shifted the climate change focal point into the Division of Operations as the director of the Policy Development and Evaluation Service did not want his staff member to spend so much time working on it. Then in 2010 the role was given to the Division of International Protection but it was still not an institutionalized full-time role with official terms of reference.[80] This continual shifting of the focal point illustrates that the agency struggled to decide how to address the issue, where it should be located within the agency and who should be working on it.

Subsequently, in 2011 UNHCR established a new position, "Senior Technical Advisor on Climate Change," to develop disaster risk reduction work within the agency. However, the position was only for one year and filled by the NRC, who paid the bulk of the advisor's salary and benefits.[81] Moreover, UNHCR sidelined this operational work and focused on the legal protection of climate change and displacement.[82] The climate change advisor investigated funding for UNHCR projects from the various climate change funds but UNHCR was not able to identify or access these successfully.[83] This was in part because there was no "buy-in" from the organization.

In summary, UNHCR did not create and fund any full time positions dedicated to working on climate change and displacement. The office made minor structural changes—establishing the climate change focal point—but no full-time permanent positions were created. Although several staff took on full-time responsibility for climate change, they were in short-term roles and not permanent positions with climate change in their terms of reference.[84] Climate change was not deemed relevant enough to their core mandate to warrant significant human resource investment.

Policy change

In mid-2008 the IASC, a coordination mechanism for humanitarian organizations based in Geneva, established an informal task force on climate change. The aim of this group was to coordinate a common humanitarian advocacy strategy on climate change and they wrote collective submissions to the annual UNFCCC COPs.[85] As will be discussed in the next chapter, IOM encouraged the creation of a special sub-group on migration, displacement, and climate change within this task force.[86] The UNHCR climate change focal point represented UNHCR at these two working group meetings, even when UNHCR had not developed a policy position in house beyond a critique of the term "climate refugee."

The IASC and Guterres's frequent rhetorical statements on climate change were a catalyst for policy development in UNHCR. In September 2008 the Policy Development and Evaluation Service published the agency's first policy paper on climate change—a year after Guterres had first raised it to the ExCom.[87] The paper was directed as much at an internal audience as an external one, and aimed to offer a preliminary policy stance and support the High Commissioner's previous rhetorical statements. The policy paper argued strongly against use of the term "environmental or climate change refugee," which it posited was a misleading and "potentially dangerous" term. It emphasized that the term "refugee" should only be applied to those covered by the

Refugee Convention. Again, UNHCR's stance reflected the agency trying to defend its core mandate.

In addition, the policy paper noted that climate change could lead to displacement but this would predominantly be internal displacement. UNHCR had responsibility for the protection of internally displaced peoples under the cluster approach, the framework for division of labour between humanitarian agencies. As it stated: "Some movements prompted by climate change could indeed fall within the traditional refugee law framework bringing them within the ambit of international or regional refugee instruments … as well as within UNHCR's framework."[88] Finally, the paper identified gaps in the international protection framework as those displaced across national borders by climate change would not fall under its mandate. The paper stated that "UNHCR does recognize that there are indeed certain groups of migrants, currently falling outside of the scope of international protection, who are in need of humanitarian and/or other forms of assistance."[89] The paper displayed a cautious shift as UNHCR acknowledged there were protection issues beyond the agency's mandate. However, it offered little insight into if and how the agency would respond to those who fell outside its mandate: the agency was sticking strongly to its mandate.

In 2009, in the lead-up to Copenhagen, UNHCR also updated the policy paper, yet it barely changed at all. The policy paper claimed that UNHCR would, going forward, have an "overarching policy to tackle the effects of climate change," which would be reflected in "operations management; protection strategies; and advocacy."[90] While these constituted bold statements there is little evidence that they were supported by any larger climate change policy development. There was, for example, no "climate strategy," although one is mentioned in the 2009 paper. Moreover, a number of UNHCR staff expressed their confusion about why UNHCR attempted to update this paper and yet made no substantive changes.[91] There was little policy development between 2008 and 2009 in UNHCR, despite the highly charged context in the lead-up to Copenhagen, the biggest climate change negotiations ever.

Meanwhile, the IASC working group became the main focus for UNHCR's climate change focal point. Initially, UNHCR was apprehensive of the working group. It had an ongoing debate with IOM over the title and terminology that should be used to refer to people who migrated or were displaced due to climate change.[92] UNHCR's position in the group was that climate change could not produce refugees in the legal or official sense and the working group was reportedly mired in definitional debates.[93] According to one IASC participant, UNHCR "argued there were existing mechanisms that could be used [to offer protection] and wanted to avoid using the term climate change refugees."[94] UNHCR's position was manifest in the first sub-working submission to the UNFCCC, *Climate Change and Migration: Who Will Be Affected?* in October 2008. This submission included a detailed description of who constituted an IDP, refugee, or a stateless person under international law and was clearly authored by UNHCR.[95] UNHCR's initial engagement in the IASC illustrated how the agency sought to defend the unique status of refugees in the face of new groups of displaced peoples.

Over 2009 UNHCR developed a more detailed policy position on climate change and statelessness and climate change displacement through the IASC. The regular meetings, with a set goal (writing submissions), and with a clear deadline in sight (the UNFCCC summit in December 2009), encouraged policy development. UNHCR worked on several other UNFCCC submissions with IOM and the NRC and became more interested in finding solutions for those displaced across borders by climate change.[96] In February 2009 UNHCR coauthored a submission that argued for the inclusion of migration and displacement in the UNFCCC final agreement. In a June 2009 submission it reiterated the core message that the term "climate refugee" should not be used and also recommended that a final UNFCCC agreement should "acknowledge the need to identify modalities of interstate cooperation to respond to the needs of affected populations who either cross an international frontier as a result of or find themselves abroad and are unable to return due to the effects of climate change."[97]

In addition, UNHCR authored a paper on statelessness and climate change. It explained that while states were likely to be "uninhabitable" long before "their full submersion," populations and governments would be externally displaced and potentially become "stateless."[98] UNHCR argued that in this event any stateless people would come under their mandate and "UNHCR would be pleased to support efforts by States to devise appropriate solutions for potentially affected populations, in partnership with other actors."[99] Statelessness fitted well within UNHCR's mandate and this was thus a much easier argument to make.

In summary, UNHCR made no policy change between 2000 and 2006 but when the issue-linkage between climate change and migration became stronger (in 2008 and 2009) the office began to publish policy papers on the issue and a series of submissions to the UNFCCC on climate change, displacement, and statelessness (2008–2009). However, these policies were not accompanied by other high-level strategic documents. Although the 2009 policy paper referred to a climate change strategy and stated that UNHCR would establish an "overarching policy to tackle the effects of climate change," this never came to fruition. Despite this lack of high-level policy, climate change was identified as an important priority in 2011 for the 60th anniversary of the Refugee Convention and the 50th anniversary of the Statelessness Convention, as will be discussed shortly.

Operational change in Kenya?

In 2013 at an international conference on climate change induced displacement, UNHCR stated that the "combination of drought and famine in the Horn of Africa in 2011 and 2012" led to a "massive influx of Somalis into Kenya's Dadaab refugee camp."[100] They saw this as an example of the "deadly interplay of deteriorating environmental conditions, political instability and conflict."[101] This crisis was framed by OCHA, and at times by UNHCR, as a climate change driven crisis, interwoven with the ongoing civil war in Somalia. Previously, UNHCR Geneva had also highlighted at the UNFCCC summit in December 2009 that climate change displacement was occurring into Kenyan refugee camps. This is one of the rationales for an examination of UNHCR's operations in Kenyan refugee camps. UNHCR have explicitly highlighted only two other examples of climate change displacement: in the Pacific Islands and in Bangladesh.

Furthermore, UNHCR Geneva published a press release stating that many Ethiopians and Somalis were being "forced to flee due to climate change and general insecurity."[102] It cited the example of an Ethiopian (but ethnic Somali), Dulane Jama, and his family who were unable to sustain their pastoralist livelihoods due to a lack of rain and thus pasture. The article maintained that Jama saw the lack of rain and weather conditions as "the root of his problem." Although most of the refugees in Dadaab had "fled conflict or persecution in their troubled homeland" (thus making them traditional refugees), Jama was "slightly different—he and his family have been forced to flee by climate change and general insecurity." This was a highly controversial and potentially dangerous statement as it undermined Jama's claim to refugee status. The article did not stipulate whether UNHCR had provided Jama and his family assistance and/or if they had gained refugee status.

However, UNHCR Kenya staff did not view climate change induced displacement as an issue of concern in the Kakuma or Dadaab refugee camps.[103] The prevailing view was that those hosted in Kenyan refugee camps had been forced to flee due to well-founded fears of persecution and fit within the convention's definition of a refugee and the

Organization for African Unity (OAU) definition.[104] None of the refugee status determination officers in Kakuma and Dadaab reported that climate change (or even proxy issues such as environmental stress or drought) had specifically been given as a cause of displacement by an interviewee.[105] Staff explained that Somali asylum seekers, who constitute over 90 percent of Kenya's refugee intake, have *prima facie* refugee status due to the prevalence of conflict, are recognized collectively as refugees, and do not need individual status determination. Thus the key component of their flight from Somalia may be drought but UNHCR is still mandated to offer them protection. The head of UNHCR's Dadaab suboffice summarized the situation as follows: "Somalis have *prima facie* status so it [climate change induced displacement] is not an issue anyway that needs to be worked through."[106] Somalis' prima facie refugee status and UNHCR's operational focus on convention refugees meant UNHCR Kenya did not perceive climate change as driving displacement.

In fact, UNHCR Kenya staff were not aware of the global debate on climate change induced displacement, had not seen UNHCR's policy briefs on the topic, and did not conceive of a "gap" in the international protection framework.[107] The head of the Kenyan Department of Refugees Affairs also claimed that there has been "no discussion with UNHCR on climate change and displacement [in Kenya]. They are full to the brim with responsibilities and won't accept any more."[108]

Yet, staff in UNHCR Kenya acknowledged that during the drought and resulting famine in Somalia there was an increase in refugee arrivals.[109] UNHCR staff in Dadaab, including the head of the suboffice, claimed there had been an increase in new arrivals when Somalia was in drought.[110] Moreover many staff highlighted the interconnections between conflict, drought, and famine leading to displacement. A UNHCR officer explained that "there are higher malnutrition rates [in Somalia] where there is drought, and where it is Al Shabaab controlled" suggesting that the presence of famine is linked to the Somali civil war.[111]

In fact, most UNHCR staff acknowledged that displacement into Kenya from Somalia was often a product of drought and conflict.[112] One staff member stated that "Somalis are coming into Kenya partly because of drought on the Somali side" and added that Somalia "does not have the capacity to host those people suffering from drought."[113] Another UNHCR staff member maintained that "the climate makes things worse. It is not the first reason why people flee but it aggravates the whole situation, even people who could find a reason to return are reluctant."[114] However, no staff used the term "climate change induced displacement" to describe such movement. They recognized that displacement had to be primarily driven by conflict for UNHCR to have a mandate and for UNHCR to offer protection.

UNHCR staff in Kenya viewed Somali asylum-seekers as traditional refugees—even if they were affected by climate change. UNHCR staff had no interest in portraying them otherwise because then they would not be offered protection. This example demonstrates how global policy developments on climate change and displacement did not resonate on the ground. Arguably a more suitable case would be climate change and displacement in the Pacific Islands, where climate change is having some of the most dramatic effects on low-lying islands. One Kiribati man has even claimed refugee status based on climate change, however this was not granted by the courts in New Zealand.[115] Regardless, UNHCR has limited presence in the Pacific and there have been no cases of statelessness for it to deal with yet. Most importantly, UNHCR has not identified any population as "climate change displaced" or sought to offer assistance on this basis. This may demonstrate how many of those affected by climate change are already captured under existing operations of UNHCR, namely through IDP assistance or conventional refugee protection (such as the Somali case). This begs a broader question: Does climate change induced displacement indeed constitute a new form of displacement which necessitates new forms of protection and assistance? UNHCR took up this question with member states in the 2010s as we will now examine.

Mandate change?

In 2011 UNHCR commemorated the 60th anniversary of the Refugee Convention and the 50th anniversary of the 1961 Statelessness Convention. The High Commissioner made climate change a core theme of the year's celebrations, which culminated in two major ministerial meetings in December 2011. UNHCR also encouraged Norway to host a conference on Climate Change and Displacement in Oslo in commemoration of Fridtjof Nansen, the first High Commissioner for Refugees. The office set the agenda by organizing a three-day expert roundtable meeting on Climate Change and Displacement: Identifying Gaps and Responses at Bellagio, Italy in February 2011. A small group of experts from academia, NGOs, and international organizations met with senior UNHCR staff and several governments to discuss the limitations of international protection frameworks for those displaced by climate change. The most significant outcome of this meeting was the recommendation that "states in conjunction with

UNHCR develop a guiding framework or instrument" to address the protection needs of those displaced by sudden-onset disasters.[116] UNHCR began 2011 with an explicit goal: to develop a new protection mandate for climate change displacement.

In May 2011 UNHCR urged states to explicitly pledge to address this protection gap at the December ministerial meeting. The agency wrote a guidance note for the pledging process in preparation for the ministerial summits. In the last section of this guidance note they encouraged states to:

> identify situations that fell outside of the scope of the existing refugee protection instruments, develop the international protection regime in a way that provides appropriate and consistent responses to these situations; and/or developing a guiding framework for temporary or interim protection scenarios identifying the circumstances in which protection would be activated, the treatment that would be provided and how it would come to an end.[117]

The pledging document did not explicitly mention "climate change induced displacement," as UNHCR was concerned that states would be less likely to support or make pledges on this agenda item if they did so. However, UNHCR hoped that states would follow the recommendations from Bellagio and be favorable to an involvement of UNHCR in addressing protection gaps related to cross-border displacement as a result of natural disaster and climate change induced displacement.

UNHCR also used the Norwegian Nansen Conference on Climate Change and Displacement to lobby states for mandate expansion. The conference in Oslo included several member state representatives (Finland, Kenya, New Zealand, Norway, and the United States), academics, NGOs, and policy makers. The High Commissioner in his opening speech outlined a role for UNHCR in offering protection to those displaced by climate change: "Even if they are not refugees, such people are entitled to our support and to have their voices heard and taken into account."[118] He called for a new legal framework for those displaced externally: "I strongly believe that a more viable approach would be to at least develop a global guiding framework for situations of cross-border displacement resulting from climate change and natural disasters." Finally he urged conference attendees to endorse the Nansen Principles as these would make a significant contribution to the 60th anniversary.

The Nansen Principles restated the Bellagio recommendation for "a more coherent and consistent approach at the international level … to meet the protection needs of people displaced externally owing to sudden onset events."[119] In addition, it called upon states "working in conjunction with UNHCR" to "develop a guiding framework or instrument in this regard."[120] What was important was that UNHCR could now claim that they had a state's (Norway) support for this and thus move the debate forward. The Nansen Principles were authored by a small group: the NRC, UNHCR, and Norwegian government representatives but the primary authorship was clearly UNHCR's. UNHCR wanted states to endorse its position as the leader and facilitator of discussions on a new protection framework.

States were suspicious of this strategy and generally unsupportive of UNHCR expansion. The Kenyan commissioner for refugee affairs, for instance, who was present at the Nansen conference, maintained that "UNHCR has already expanded to include statelessness, IDPs and [now] including environment-related movers, a fourth category of people, how amorphous is this organization going to be?"[121] He was against mandate expansion: "If [financial] support for UNHCR stays the same then they should leave it to another agency so that resources aren't spread too thinly."[122] Some states argued that it was "too early" to talk about developing soft-law frameworks for climate change displacement. Other states expressed concern that UNHCR did not have capacity or financial resources to expand particularly given that they had enough difficulty fulfilling their obligations to refugees.

Nevertheless, the High Commissioner continued to campaign for a new protection framework. In November 2011 Guterres made a speech to the UN Security Council on climate change and displacement. He stated, "More and more people are being forced to flee due to reasons that are not covered by the 1951 Refugee Convention."[123] He argued that it was a "humanitarian imperative" to assist those displaced by climate change or other natural disasters. Finally, he recommended that the international community "formulate and adopt a set of principles, specifically designed to reinforce the protection of and find solutions for people who have been forced to leave their own country as a result of catastrophic environmental events, and who may not qualify for refugee status under international law." In this speech he promoted expansion over affirming UNHCR's core identity. Although he attempted to link this call for a new protection framework to UNHCR's existing responsibilities, these issue-linkages were tenuous. He reiterated this demand for a new international protection framework to states at the December ministerial meeting and stated that UNHCR was "ready to work with states who want to help develop such guiding frameworks." UNHCR's emphasis on expanding the international protection framework reflects how staff saw their organization's mandate and expertise.

The December ministerial meeting was a disappointment for UNHCR as it did not result in a new mandate for climate change displacement. UNHCR prioritized "future protection challenges" and burden-sharing as the two major issues for the ministerial meeting. In the UNHCR background paper they highlighted the gap in the international normative framework, and recommended a new "global guiding framework" or instrument for displaced peoples who did not fit inside the convention. This guiding framework was primarily aimed at climate change and natural disaster related displacement but UNHCR asked states if it could be broader. However, only five states (Argentina, Germany, Mexico, Norway, and Switzerland) pledged to develop a new protection framework and there was no mention of climate change or natural disaster displacement in the final ministerial communiqué. UNHCR did not gain a mandated expansion of its activities in 2011.[124] However UNHCR had changed its rhetoric and policy position significantly but not its structure on the issue of climate change and displacement.

Despite this lack of state support, UNHCR continued to work on solutions for those displaced across borders by climate change. In particular, staff worked with Norway on the Nansen Initiative, a continuation of the agenda set out at the Nansen Conference in Oslo in 2011. The creation of the initiative was spurred by the ministerial conference where Norway and Switzerland were two of just five states that pledged support for developing new protection frameworks for those displaced across international borders due to climate change and natural disasters. The Nansen Initiative aims to build consensus amongst states on a global protection agenda for "disaster-induced cross border displacement" and is run by a steering group of states (Australia, Bangladesh, Costa Rica, Germany, Kenya, Mexico, Norway, Philippines, and Switzerland), and chaired and funded by Norway and Switzerland. It also has a small secretariat in Geneva and an envoy, Walter Kalin, who was the former United Nations Special Representative for the Human Rights of Internally Displaced Persons. However, the initiative has clearly stated it does not intend to follow the IDP process of developing a "soft-law instrument" to protect those displaced internationally by climate change.

Rather, it seeks to build consensus on local operational responses through five regional consultations in areas likely to have "disaster-induced cross-border displacement," and commission research into the phenomena, and then feed the lessons learned back into key global summits including the UNFCCC 2015 Summit in Paris, as well as 65th and 66th UNHCR ministerial meetings.[125] UNHCR's involvement has been central to the Nansen Initiative. The European Commission has given a collective grant to UNHCR, the NRC, and the Nansen Initiative to commission research studies, convene the five regional and intergovernmental consultations, and disseminate the outcomes of the Nansen Initiative process. More recently, in March 2014 UNHCR and the Nansen Initiative organized a conference of experts on Planned Relocations within national borders at San Remo, Italy with the Brookings Institute.[126]

Outside of the Nansen Initiative, UNHCR has also continued to speak out on the protection needs of those displaced by climate change. In 2012 the world's focus was on Rio+20: the 20-year anniversary conference of the UNCED. Guterres attended these negotiations and reiterated the message of a report on environmental displacement in East Africa. He emphasized that "this report confirms what we have been hearing for years from refugees. They did everything they could to stay at home, but when their last crops failed, their livestock died, they had no option but to move; movement which often led them into greater harm's way."[127] However the final Rio+20 agreement had no mention of displaced persons or refugees, despite Guterres's efforts.

Other senior staff have also made clear UNCHR's commitments to work on climate change displacement. In 2013, José Riera, the senior advisor to the director of International Protection, set out the problem at a summit in Berlin on climate change displacement. He noted there are likely to be more people displaced due to climate change, but these are primarily the concern for national governments if they are IDPs. However, UNHCR had an "institutional interest in the topic" as he stated:

> While its [UNHCR's] refugee mandate clearly does not encompass displacement caused by natural disasters and climate change, the organization has a clear interest in the movement of people prompted by these factors. Environmental degradation can fuel social tension and, in some cases, conflict which, in turn, can give rise to flows of refugees and IDPs. Even where the cause of displacement—whether internal or cross-border—is primarily environmental, the affected populations may have protection needs and vulnerabilities similar to those whose flight is provoked by violence or human rights abuses.[128]

Then in July 2014 the director of the Division of International Protection, Volker Türk, explained at another summit in Berlin that states did not give UNHCR a clear mandate to work on climate change related displacement "but we will want to continue our work, we are not intimidated," as he explained, "of course this is a topic that we deal with."[129] This is an excellent example of how UNHCR has sought to link protection of peoples displaced across borders by climate change to its traditional refugee mandate, and demonstrates how far the UNHCR position came in the early 2010s. UNCHR has continued to explore protection for those displaced by climate change, and sought to galvanize state support for an expanded protection framework.

UNHCR and natural disasters

Parallel to these debates on climate change and displacement, there were ongoing discussions over UNHCR's protection role for IDPs, including those displaced by natural disasters.[130] Although there was substantial overlap in *who* would be affected—people affected by floods, droughts, and other extreme weather events—the policy discussions on these two areas were kept relatively separate.[131] UNHCR's engagement with natural disaster IDPs came out of broader concerns about humanitarian reform and IDPs, and was not framed as a response to climate change. There were thus two significant differences in focus in this work: 1) a focus on those displaced internally, and not internationally; and 2) a focus on a broader category of natural disasters and not just those that are linked to climate change but also those that are not (tsunamis and earthquakes).

In the 2000s UNHCR began to offer assistance on an ad-hoc basis to peoples displaced internally by natural disasters. It offered assistance for the first time in Sri Lanka and Indonesia following the Asian tsunami of 2004. It made this "unprecedented move" upon the request of the Secretary General and made clear it was doing this beyond its formal mandate.[132] The office justified their involvement due to their "heavy operational presence on the ground and the massive scale of the disaster and need for immediate humanitarian action."[133] Following this, UNHCR took action in six other natural disaster situations, including: the Pakistan earthquake (2005); cyclone Nargis in Myanmar (2008); the Philippines floods (2009); the Pakistan floods (2010); and the Haitian earthquake (2010).[134] In total, between 2005 and 2010 UNHCR had an operational involvement in thirteen out of fifty-eight natural disasters, and provided support in another five. In all of these cases UNHCR already had an operational presence in the country and thus there was a humanitarian imperative to assist. It was difficult for the organization not to offer its expertise and resources to assist the local population.[135]

UNHCR also offered assistance as a symbolic gesture of international solidarity. Finally, there were instrumental advantages to this assistance. By assisting local IDPs it could build better relations with host governments which it hoped would benefit its refugee operations in-country.[136]

UNHCR differentiated its protection mandate for those affected by conflict from those affected by natural disaster. UNHCR was the sole cluster lead at the country level for protection of conflict-induced IDPs. However, its cluster protection role for natural disaster IDPs was shared at the country level with Office of the High Commission for Human Rights and the United Nations' Children's Fund (UNICEF).[137] The agency perceived it had a critical role in protection for conflict-induced IDPs but not for natural disaster IDPs. This delineation is important as it shows the agency sought to establish a clear link to its mandate: it would protect those IDPs that could potentially become refugees (those facing persecution and conflict) but not those facing environmental disasters. In other words, UNHCR distinguished between natural disaster and conflict IDPs as the latter fitted more closely with its mandate, although the convention does not refer to conflict.

In addition, UNHCR ensured that refugees remained outside of the cluster system during the process of humanitarian reform. The agency saw refugees as a distinct group, separate from other victims of humanitarian crises, and did not want to integrate them into the humanitarian cluster system. It saw itself as the sole overseeing agency responsible for refugee matters, which ran counter to the logic of the cluster system, which allocated responsibility along thematic sectorial lines (such as water and sanitation). From an operational perspective this was highly problematic: how could one distinguish between refugees in complex emergencies where there were often many causes of movement? Furthermore creating a special exemption for refugees undermined the plan for a universal, streamlined, and predictable humanitarian response framework.

The head of the UN's humanitarian response sought to convince UNHCR to include refugees within the cluster system. In February 2009 the emergency relief coordinator, John Holmes, met with Guterres to discuss UNHCR's role in the cluster system.[138] Holmes encouraged the High Commissioner to ask the General Assembly for a broader humanitarian mandate and not one with a dominant focus on refugees, which had the effect of leaving refugee emergencies outside the cluster response. He assumed that UNHCR's refugee mandate could be easily changed by General Assembly resolution. The High Commissioner responded by explaining that the UNHCR mandate was more than a General Assembly resolution.[139] UNHCR's mandate could not be so easily changed as it was also based on the Refugee

Convention. Guterres had the 1951 Convention on the Status of Refugees on hand and was able to show it to Holmes and explain UNHCR's supervisory status over the convention.[140] This is a compelling example of how UNHCR closely protected its core mandate.

However, the High Commissioner did seek to expand UNHCR's humanitarian activities in addition to its refugee mandate over the course of the 2000s. Rather than asking the General Assembly to change UNHCR's statute, as Holmes had suggested, Guterres sought to formalize a new role for UNHCR in assisting internally displaced peoples in natural disaster situations. In September 2009 in a speech to the ExCom he proposed that UNHCR take a "more predictable role" as the protection lead at the country level in natural disasters.[141] States were not supportive of such an expansion, although they supported the High Commissioner and reelected him for a second term in office in April 2010. In speeches to UNHCR staff after his reelection he highlighted his expansionist tendencies, stating that he saw UNHCR as the lead humanitarian actor for "forced migration."[142] In line with this vision, one of his top priorities was to develop the office's ability to rapidly respond to humanitarian emergencies and take on a wider set of responsibilities within the cluster system.[143]

Over the course of 2010 and 2011 Guterres sought a more clearly defined mandate and role for UNHCR in natural disasters. He made a bid directly to the IASC heads of agencies to be the cluster lead for the protection of IDPs in natural disasters and conflict in early 2010.[144] In addition, the agency published a review arguing for UNHCR to take on a "more predictable role" in natural disasters.[145] UNHCR stated that until 2009 there had been "no institutional appetite for involvement in natural disasters, the international community's focus on climate change and natural disasters has changed that, and, it would appear, so have institutional considerations about UNHCR's future role as *the* protection agency."[146] It claimed that UNHCR's involvement with natural disasters had been "unpredictable," and "by exception and slow." This was problematic for UNHCR as it could not access flash appeal funding in natural disasters and made it difficult for other NGOs and partners who were not sure of UNHCR's role. It proposed that UNHCR should "provide leadership, especially in regard to any 'protection gaps' rising from the debate over climate change and environmental refugees."[147] It saw UNHCR's expansion into natural disaster protection as "consistent with UNHCR's protection work with IDPs to date ... and a natural progression from its role as lead of the Global Protection Cluster."[148] This policy paper was a significant policy development and illustrates how expansion into climate change and displacement overlapped with, and may have even triggered expansion into IDP natural disaster assistance.

In December 2010 the IASC endorsed a UNHCR 12-month "pilot program" as the protection coordinator at country level in natural disasters.[149] Following this, UNHCR went to its member states for a mandate for this pilot program. States resolutely rejected it. At the Standing Committee in March 2011 several delegations cautioned that UNHCR's engagement with and protection of natural disaster victims "should not be at the expense of its core mandate; that it should be subject to the consent of the Government concerned; and that it should not entail any diversion of funding from core mandate activities."[150] States were not supportive of UNHCR becoming the lead humanitarian agency for protection in natural disasters, just as they were not supportive of it expanding its legal protection mandate for internationally displaced peoples due to climate change.

Despite a lack of state support UNHCR has continued to participate in humanitarian operations to assist internally displaced peoples affected by natural disasters. It has justified this in terms of its cluster responsibilities. José Riera explained in 2013:

> UNHCR's traditional core mandate does not encompass internal displacement caused by natural disasters and climate change. By operation of the inter-agency division of labor on IDPs, known as the "cluster approach," since 2005 we have contributed to ensuring greater leadership and accountability, and a more effective and predictable inter-agency response for IDPs.[151]

UNHCR's general stated policy is to decide on a case-by-case basis whether it will be involved. It is likely to do so if it has an established presence and relief items in the country in which a disaster strikes.[152] Its involvement also relies on an invitation from the disaster-affected country, and the Emergency Relief Coordination, on the basis that no other agency has the capacity to lead. In February 2013 for instance, UNHCR was operating in Indonesia and the South Pacific, two out of eight natural disaster emergencies in which the cluster system had been activated.

Overall, UNHCR sought to informally expand its mandate and offer protection in two ways. Firstly it sought a mandate from the IASC (2010) and then states (2011) to protect natural disaster IDPs. It had not succeeded in obtaining such a mandate by December 2011. In addition, UNHCR lobbied states during the Nansen Conference and in the lead-up to the 2011 ministerial conference for a role in developing a new "framework" for those displaced internationally by climate change and natural disasters. However, this was not supported by the majority of states either. Yet UNHCR worked with those supportive states through the Nansen Initiative after 2011 to continue to lobby for an expanded protection framework. Interestingly, these two policy developments (natural disaster IDPs and climate change induced displacement) occurred somewhat separately within the organization. Although they were often linked at a rhetorical level by the High Commissioner, there was no clear policy link between them.

Yet UNHCR did expand and respond to natural disasters during this period. Throughout the 2000s it offered ad-hoc assistance to victims of natural disasters. However, these operational activities do not indicate UNHCR explicitly engaging with climate change but rather responding to new humanitarian needs. UNHCR's humanitarian activities during the 2000s suggest that there is scope for addressing internal displacement related to climate change within existing structures.

Finally, there is a mismatch between operational categories and the policy process at headquarters. While UNHCR staff have invested in developing responses to climate change displacement, no UNHCR staff could identify a population displaced internationally by climate change related disasters. For example, UNHCR has claimed in policy and rhetorical statements that statelessness on low-lying islands may be an issue, however it has not worked significantly on this issue in the Pacific or the Caribbean.[153] In sum, discussions over extending protection and assistance to "climate change induced displacement" do not appear to have much purchase beyond UNHCR headquarters, although natural disasters do.

Conclusion

UNHCR's core mandate as a refugee protection agency has shaped how staff have responded to climate change. UNHCR was initially reluctant to engage with the issue, although NGOs, academics, and other international organizations were increasingly vocal about the link between climate change and displacement in the mid- to late 2000s. This was primarily because staff argued that those affected by climate change did not fit within the 1951 Refugee Convention, a constitutional document for UNHCR. UNHCR staff made minimal policy, structural, and operational changes in the late 2000s, despite the significant public attention to climate change in the lead-up to Copenhagen in 2009.

However, the High Commissioner was more eager to expand despite state resistance. He continued to lobby states to consider expanding the protection mandate of UNHCR—in various speeches and most notably at the 2011 Nansen Conference, at the 2011 ministerial meeting, and through protection for IDPs affected by natural disasters. The Nansen Principles, coauthored by UNHCR, the NRC, and the Norwegian government in 2010, were one clear manifestation of UNHCR's intentions to extend its international protection mandate for those displaced across borders due to climate change. In parallel, UNHCR has continued to offer humanitarian assistance to countries affected by natural disasters, without labeling this explicitly as a response to climate change.

The UNHCR case demonstrates the importance of staff—and in particular the High Commissioner—for shaping mandate expansion. Staff's perception of their mandate as refugee protection and not assisting all displaced peoples explains why they were initially reluctant to engage in debates on climate change and displacement. Although states set formal parameters of the mandate and resisted expansion of the protection mandate or humanitarian assistance, the High Commissioner and UNHCR staff sought other ways to pursue a more expansive role. They collaborated with sympathetic states through the Nansen Initiative and assisted people affected by natural disasters. The UNHCR case suggests that international organizations will not always seek to expand, and staff will carefully consider how new issues relate to their core mandate. It also highlights the role that executive heads play in convincing staff and states alike to expand.

Finally, this chapter suggests that expansion can occur de facto without a formal de jure mandate from states. Although member states expressed concern with UNHCR mandate expansion repeatedly between 2007 and 2014 the High Commissioner and senior staff continued to lobby states for a mandate, and most importantly continued to find ways to work on the issue regardless. And who knows, in the future UNHCR staff may successfully change state preferences and gradually gain some formal mandate expansion.

Notes

1 Alexander Betts, Gil Loescher, and James Milner, *UNHCR: The Politics and Practice of Refugee Protection* (Abingdon: Routledge 2012).

2 Jeff Crisp, "Beyond the Nexus: UNHCR's Evolving Perspective on Refugee Protection and International Migration," *New Issues in Refugee Research* Research Paper, no. 155 (2008); Martin Gottwald, *Directive versus Facilitative Leadership in Times of Change: UNHCR's Organizational Culture and Decision-Making Processes in Light of the*

Humanitarian Reform and the Cluster Approach (Oxford: Refugee Studies Centre, University of Oxford: 2009).

3 Frank Biermann and Ingrid Boas, "Preparing for a Warmer World: Towards a Global Governance System to Protect Climate Refugees," *Global Environmental Politics* 10, no. 1 (2010): 60–88.

4 Betts, Loescher, and Milner, *UNHCR: The Politics and Practice of Refugee Protection.*

5 1951 Convention Relating to the Status of Refugees, Article 35.

6 *1950 Statute of the Office of UNHCR*, Paragraph 8(a), Chapter II.

7 Tracy Glover and Simon Russell, "Overseeing the Refugee Convention: Coordination with UNHCR and States," *ICVA Working Paper*, no. 7 (2001); Volker Türk, "UNHCR's Supervisory Responsibility," in *New Issues in Refugee Research, Working Paper No. 67* (Geneva, Switzerland: UNHCR, 2002), 1–20.

8 1951 Convention Relating to the Status of Refugees, Article 1A (2).

9 The Oxford English Dictionary defines a refugee as "A person who has been forced to leave his or her home and seek refuge elsewhere, esp. in a foreign country, from war, religious persecution, political troubles, the effects of a natural disaster, etc.; a displaced person." Oxford English Dictionary 2015, *Oxford English Dictionary*, www.oed.com. For more on definitions of refugees see: Andrew Shacknove, "Who Is a Refugee?," *Ethics* 95, no. 2 (1985): 274–284.

10 Türk, "UNHCR's Supervisory Responsibility," 499.

11 Gil Loescher, *The UNHCR and World Politics: A Perilous Path* (Oxford: Oxford University Press, 2002).

12 UNHCR, *Report of the United Nations High Commissioner for Refugees, Part II, Strategic Review Pursuant to General Assembly Resolution 58/153* (Geneva, Switzerland: UNHCR, 2013).

13 Türk, "UNHCR's Supervisory Responsibility," 497; Loescher, *The UNHCR and World Politics: A Perilious Path*, 5.

14 Betts, Loescher, and Milner, *UNHCR: The Politics and Practice of Refugee Protection.*

15 Türk, "UNHCR's Supervisory Responsibility," 481.

16 Ann Mckittrick, "UNHCR as an Autonomous Organisation: Complex Operations and the Case of Kosovo," *RSC Working Paper* 12, no. 50 (2008).

17 Michael N. Barnett and Martha Finnemore, *Rules for the World, International Organizations in Global Politics* (Ithaca, NY: Cornell University Press, 2004), 725.

18 UNHCR had also inherited significant moral legitimacy in this realm from its other predecessor: the League of Nations' Office of the High Commissioner of Refugees, led by Fridtjof Nansen, a charismatic Norwegian diplomat, explorer, and humanitarian, who had a high degree of moral authority. Loescher, *The UNHCR and World Politics: A Perilous Path.* On the early refugee regime see also: Guy Goodwin-Gill and Jane McAdam, *The Refugee in International Law* (Oxford: Oxford University Press, 2007).

19 Betts, Loescher, and Milner, *UNHCR: The Politics and Practice of Refugee Protection*, 116.

20 Betts, Loescher, and Milner, *UNHCR: The Politics and Practice of Refugee Protection*, 74.
21 Betts, Loescher, and Milner, *UNHCR: The Politics and Practice of Refugee Protection*, 83.
22 Interview with UNHCR member state, 10 May 2011, Geneva.
23 UNHCR's 1950 statute and the 1951 Refugee Convention limited UNHCR to work with refugees displaced before January 1951. See: *1950 Statute of the Office of UNHCR*, Paragraph 6 (a) (i) and 1951 Convention Relating to the Status of Refugees.
24 Loescher, *The UNHCR and World Politics: A Perilous Path*.
25 *1950 Statute of the Office of UNHCR*, Paragraph 1 (a).
26 *1967 Protocol Relating to the Status of Refugees*.
27 UNHCR had a mandate to examine and assist individual stateless persons from the 1954 and 1961 conventions. It gained a global mandate "to prevent and reduce statelessness" from the 1994 General Assembly resolution 49/169. For a more detailed discussion of UNCHR's stateless mandate see: UNHCR, *Extracts Relating to UNHCR's Supervisory Responsibility for the Statelessness Conventions*, www.unhcr.org/refworld/docid/4f5cd0692.html; UNCHR, "Stateless—UNCHR Actions," www.unhcr.org/pages/49c3646c16a.html; and UNHCR—Division of International Protection 2010, *UNHCR Action to Address Statelessness*, www.unhcr.org/4b960ae99.html.
28 The staff and budget figures for UNHCR can be found online at: UNHCR 2015, "Staff Figures," www.unhcr.org/pages/49c3646c17.html; and UNHCR 2015, "Financial figures," www.unhcr.org/pages/49c3646c1a.html.
29 Raquel Freitas, "UNHCR Decisions on Internally Displaced Persons," in *Decision Making within International Organizations*, ed. Bob Reinalda and Bertjan Verbeek (London: Routledge, 2004), 123–136.
30 *1950 Statute of the Office of UNHCR*, Paragraph 8(a), Chapter II.
31 Freitas,"UNHCR Decisions on Internally Displaced Persons."
32 Guy Goodwin-Gill, "UNHCR and Internal Displacement," *World refugee Survey 2000* (2000): 26–31.
33 Ibid.
34 Betts, Loescher, and Milner, *UNHCR: The Politics and Practice of Refugee Protection*, 113–114.
35 Freitas,"UNHCR Decisions on Internally Displaced Persons."
36 Anonymous, "The UNHCR Note on International Protection You Won't See," *International Refugee Law* 9 (1997): 267–273.
37 Ibid.
38 ExCom was created in 1958 with 25 members and has expanded progressively and, as of September 2012, had 87 members. It meets annually to approve UNHCR's budget and program for the following year and provide guidance on UNHCR's management, objectives and priorities; see Betts, Loescher, and Milner, *UNHCR: The Politics and Practice of Refugee Protection*, 77; and UNHCR, "Executive Committee," www.unhcr.org/pages/49c3646c83.html.
39 Loescher, *The UNHCR and World Politics: A Perilous Path*.
40 Betts, Loescher, and Milner, *UNHCR: The Politics and Practice of Refugee Protection*, 93.

41 Interview with UNHCR member state representative, 10 May 2012, Geneva.

42 Telephone interview with UNHCR member state, 19 June 2012.

43 Loescher, *The UNHCR and World Politics: A Perilous Path.*

44 Human-made disasters were defined as "international or internal armed conflict, massive violations of human rights and industrial accidents." Broadly speaking, this category was seen to encompass political, social and economic factors. UNHCR, *Working Group in 1991 on Solutions and Protection to the 42nd Session of the Executive Committee of the High Commissioner's Programme* (EC/SCP/64), 12 August 1991.

45 Ibid., 11.

46 UNHCR cohosted a symposium with IOM on environment and migration. However UNHCR's primary focus was on the impact of refugees on the environment. IOM and UNHCR. 1996. *Environmentally-Induced Population Displacements and Environmental Impacts Resulting from Mass Migrations,* Geneva.

47 Ibid.

48 UNHCR, "The Environment—A Critical Time", *Refugees Magazine No. 127*, (Geneva, Switzerland: UNHCR, 2002), 2.

49 IOM, *Migration and the Environment* (Geneva, Switzerland: IOM, 1992), 48.

50 The division received much support from the Japanese government who paid the salaries of a number of the environmental advisors. Interview with UNHCR official, 17 March 2010, Geneva.

51 UNHCR 2009, "Environmental Publications," www.unhcr.org/3b94c8364.html.

52 Some of these environmental impact activities were later framed as mitigating climate change. UNHCR Kenya, for instance, hired an environmental advisor in the Kakuma refugee camp to develop a renewable energy project, funded by Energies of Portugal. Interview with UNHCR Kenya official, 5 April 2011, Nairobi. See also Energies of Portugal, www.edp.pt/en/sustentabilidade/sociedadeecultura/IntervencaoSocial/Pages/Projectokakuma.aspx.

53 Telephone interview with UNHCR senior official, 14 May 2010.

54 Phone interview with director of the Red Cross/Red Crescent Climate Change Center, 16 April 2010. Also interviews with other IASC members March–May 2010, Geneva.

55 Carol Farbotko, "Tuvalu and Climate Change: Constructions of Environmental Displacement in the Sydney Morning Herald." *Geografiska Annaler* 2005, 87(B): 279–293. The term "climate refugee" was predominantly used by environmental NGOs and media who were not versed in refugee law.

56 Interview with UNHCR Official, 11 November 2011, Sydney.

57 Telephone interview with UNHCR senior official, 14 May 2010.

58 Jane McAdam, "'Disappearing States', Statelessness and the Boundaries of International Law," in *Climate Change and Displacement: Multidisciplinary Perspectives,* ed. Jane McAdam (Oxford: Hart Publishing, 2010)

59 The staff member claimed there were two types of people within UNHCR: those that took a "practical approach" and talked about climate change and lawyers who tended to reject the concept. He caveated this by stating that there were some lawyers who saw the conversation

about climate change leading to statelessness as an interesting—albeit intellectual—conversation. Telephone interview with UNHCR senior official, 14 May 2010.

60 In a letter, written soon after Guterres was short-listed for the role of High Commissioner, he acknowledged the core identity of UNHCR stating: "We are an organization for refugees; let's not leave any doubt about it." He also claimed that if elected he would "strongly concentrate in creating the conditions to mobilize the entire organization around one single, but consistent and comprehensive, concept of protection and the capacity to operationalize it, making use of the 'strong leadership to instill such a culture of protection.'" António Guterres, *Talk Back, Special Issue: UNHCR Candidates Talk Back*, 2005, 7–2a.

61 He also highlighted the impact of climate change on displacement at subsequent executive committee meetings in 2008 and 2009. See: Statement by the High Commissioner, *Executive Committee of the Programme of the United Nations High Commissioner for Refugees*, Fifty-eighth Session, Geneva, 1 October 2007.

62 Interview with UNHCR official, Geneva, 17 March 2010.

63 Jeff Crisp, "Refugees, Persons of Concern, and People on the Move: The Broadening Boundaries of UNHCR," *Refuge* 26, no. 1 (2009): 73–76. Note that Crisp's views are his own, and not expressing an official UNHCR position.

64 Interview with UNHCR member state representatives, 7 and 10 May 2012. Telephone interview with UNHCR member state representative, 19 June 2012.

65 See António Guterres, "Millions Uprooted: Saving Refugees and the Displaced," *Foreign Affairs* 87, no. 5 (2008): 90–100. The Assistant High Commissioner, Craig Johnston, was the first senior UNHCR manager to attend a UNFCCC meeting. He attended the Poznan summit in 2008 at his own initiative. Interview with former UNHCR senior official, 27 January 2012, Oxford. He outlined that UNHCR did not have a mandate to protect people displaced by natural disaster but have been called upon to do so time and time again. In his words, the UNHCR mandate "related to refugees and the protection, shelter and management of internally displaced people. We are not mandated to come to the assistance of the victims of natural disasters but time and time again we have been called upon to play this role, not because it is an implied mandate but because in the face of humanitarian suffering those who can help are those who receive the call." Remarks by Craig Johnston, UN Deputy High Commissioner for Refugees, *Planning for the Inevitable, the Humanitarian Consequences of Climate Change*, UNFCCC COP 14, 2008, Poznan.

66 António Guterres, 2009, "Five 'Mega-Trends'—Including Population Growth, Urbanization, Climate Change—Make Contemporary Displacement Increasingly Complex, Third Committee Told," Third Committee, *General Assembly* GA/SHC/3964, New York: United Nations, 3 November 2009, and Statement by Mr. António Guterres, 2011, "Maintenance of International Peace and Security: New Challenges to International Peace and Security and Conflict Prevention," *United Nations Security Council Briefing*, New York: United Nations, 23 November 2011.

67 Julian Borger, "Climate Change Refugees: 'Nature's Retaliation against Human Aggression'," *The Guardian*, 17 June 2008.
68 António Guterres, "Statement by António Guterres," Nansen Conference on Climate Change and Displacement, Oslo, 2011.
69 UN General Assembly, Executive Committee of the Programme of the United Nations High Commissioner for Refugees, 60th session, summary record of the 631th meeting (A/AC.96/SR).
70 Interview with UNHCR member state representative, Geneva, 23 March 2010.
71 This was the only occasion that the High Commissioner used the term "climate refugees." He made the statement in French, see Grégoire Allix, "António Guterres 'La distinction entre réfugiés et déplacés est dépassée,'" *Le Monde*, 16 December 2009.
72 Interview with UNHCR official, 17 March 2010, Geneva. Telephone interview with UNHCR official, 30 March 2010.
73 Interview with former senior UNHCR staff member, 27 January 2012, Oxford.
74 Ban Ki-moon, *Letter to UN Agencies*, May 2008.
75 UNHCR, *Task Force on Climate Change and Related Issues, Final TORs [Terms of Reference]*, August 2008 (Geneva, Switzerland: UNHCR, 2008). On file with the author.
76 Telephone interview with UNHCR official, 30 March 2010. Interview with UNHCR official, 17 March, Geneva.
77 Telephone interview with UNHCR official, 30 March 2010.
78 Interview with UNHCR senior official, 10 May 2012, Geneva.
79 Telephone interview with UNHCR official, 30 March 2010.
80 In fact, the original climate change focal point from 2008 took over the role. Telephone interview with UNHCR senior staff member, 14 May 2010.
81 Interview with UNHCR official, 6 June 2011, Oslo.
82 Interview with UNHCR official, 9 May 2012, Geneva.
83 Ibid. Note that in 2010 UNHCR wrote a report exploring options for financing from the adaptation financing mechanisms. See UNHCR, "Climate Change, Adaptation Financing Mechanisms and Their Implications for UNHCR Operations." Division of Support and Operational Management, 31 March 2010.
84 There were only two UNHCR staff to work on the issue full-time and both did so for a short period of time between missions and their positions were not institutionalized in the UNHCR organigram. Interview with former UNHCR senior official, 27 January 2012, Oxford.
85 Interview with IASC members, March 2010, Geneva.
86 Telephone interview with IASC member, 14 April 2010. For more on the IASC see: IASC, http://interagencystandingcommittee.org/.
87 UNHCR, *Climate Change, Natural Disasters and Human Displacement* (Geneva, Switzerland: UNHCR, 2008).
88 UNHCR, *Climate Change, Natural Disasters and Human Displacement*, 11.
89 UNHCR, *Climate Change, Natural Disasters and Human Displacement*, 9.
90 UNHCR, *Climate Change, Natural Disasters and Human Displacement* (Geneva, Switzerland: UNHCR, 2009).
91 Interview with UNHCR staff members March–May 2010.

92 Telephone interview with IASC member, 14 April 2010.

93 Comments were made on this by a number of IASC members including: telephone interview with IASC member, 14 April 2010 and OCHA official, 20 March 2010, Geneva.

94 OCHA staff member, 20 March 2010, Geneva.

95 Informal group on Migration/Displacement and Climate Change of the IASC, *Climate Change, Migration and Displacement: Who Will Be Affected?* Working Paper, 31 October 2008.

96 IOM, UNHCR and UNU, *Climate Change, Migration, and Displacement: Impacts, Vulnerability, and Adaptation Options*, Submission to the 5th session of the Ad Hoc Working Group on Long-Term Cooperative Action under the Convention (AWG-LCA 5). Bonn, 29 March–8 April 2009.

97 IASC Working Group on Climate Change, *Comments and Proposed Revisions to the Negotiating Text Prepared by the Chair of the UNFCCC Ad Hoc Working Group on Long-Term Cooperative Action,* June 2009.

98 UNHCR, *Submission Climate Change and Statelessness: An Overview*, 15 May 2009.

99 Ibid., 2.

100 José Riera, "Challenges Relating to Climate Change Induced Displacement." Remarks by José Riera, International Conference "Millions of People without Protection: Climate Change Induced Displacement in Developing Countries" Berlin, 29 January 2013.

101 José Riera, "Challenges Relating to Climate Change Induced Displacement." Remarks by José Riera, International Conference "Millions of People without Protection: Climate Change Induced Displacement in Developing Countries," Berlin, 29 January 2013.

102 Andy Needham 2009, "Nowhere to Hide from Climate Change in Kenyan Refugee Camp," www.unhcr.org/4b2b76a79.html.

103 The author visited the Dadaab camps (the largest refugee camp in the world at the time) and the Kakuma refugee camps in April 2011. UNHCR does ask new arrivals at Kenyan camps their "primary reason of flight" but this data is of very questionable validity. The office sorts responses into the following categories: "access to food/assistance," "livelihood problems (environmental)," as well as "attack by fighting forces," "political reasons/threats" and "general insecurity." According to these registration statistics in 2010 the vast majority of new arrivals to Dadaab reported they had fled due to general insecurity (85.56 percent). The second most influential reason was food insecurity (2.76 percent); followed by livelihood problems—environmental (2.33 percent) and livelihood problems—security (2 percent). These categories are highly problematic as they overlap—the distinction between food insecurity and livelihood problems (environmental) is not clear for instance. Interview with UNHCR official, 21 April 2011, Dadaab; UNHCR Flight Reasons Statistics between 1 January 2010 and 31 December 2010.

104 Interviews with UNHCR staff in Nairobi, Dadaab, and Kakuma, March–April 2011. Note that the definition of a refugee in Africa is based on the OAU definition and thus broader than the UNHCR convention definition. As mentioned previously in chapter two, the 1969 OAU defines a refugee as any person compelled to leave his or her

country "owing to external aggression, occupation, foreign domination or events seriously disturbing public order in either part or the whole of his country or origin or nationality." 1989 Convention Governing the Specific Aspects of Refugee Problems in Africa, Article 1 (2), 10 September 1969.

105 Interviews with UNHCR Refugee status determination officer, April 2011, Dadaab and Kakuma. Although, refugees may not cite drought or general environmental insecurity as a reason for flight as they are often taught by others what to say to gain refugee status. One UNHCR official described Somalis as "professional refugees" suggesting that they know what to say to gain refugee status. Interview with UNHCR Head of Dadaab Sub-Office, 20 April 2011, Dadaab.

106 Interview with UNHCR Head of Sub-Office, 20 April 2011, Dadaab.

107 Interviews with UNHCR officials in Kakuma (7–15 April) and Dadaab (20–21 April).

108 Interview with Kenyan Commissioner for Refugee Affairs, 7 June 2011, Oslo.

109 This trend became particularly noticeable during the famine and humanitarian crisis in Somalia between July and November 2011, which saw thousands of new arrivals to Dadaab. However it is important to note that drought cannot be directly attributed to climate change, nor can any other extreme weather events.

110 Interview with UNHCR Head of Dadaab Sub-Office, 20 April 2011, Dadaab.

111 Interview with UNCHR officer, 21 April 2011, Dadaab.

112 I talked with a number of refugees during my visit to Dadaab and Kakuma but did not have the opportunity nor the resources to conduct a rigorous, systematic survey of causes of movement. Anecdotally there was a mixed response from refugees across the camps and destination countries—some refugees highlighted conflict as a reason for their displacement, and some stated that drought was an issue.

113 Interview with UNHCR official, 29 March 2011, Nairobi. A NRC staff member explained that "drought has diminished the livelihoods of people" and "they have nothing else if their livelihoods are destroyed." Interview with NRC official, 19 March 2011, Nairobi.

114 Interview with UNHCR official, 5 April 2011, Nairobi.

115 Morgan Godfery, "New Zealand Refuses Climate Change Refugees—Mass Action Is Now Needed," *The Guardian*, 11 May 2014. www.theguardian.com/commentisfree/2014/may/12/new-zealand-refuses-climate-change-refugees-mass-action-is-now-needed.

116 "Chair Person's Summary," *Nansen Conference on Climate Change and Displacement in the 21st Century,* Oslo, 6–7 June 2011. www.regjeringen.no/globalassets/upload/ud/vedlegg/hum/nansen_summary.pdf

117 UNHCR, 2011, *Additional Guidance Note to Support the State Pledges Process* (Geneva, Switzerland: UNHCR, 26 May 2011), 9.

118 UNHCR, "Statement by António Guterres", *Nansen Conference on Climate Change and Displacement*, Oslo, 6 June 2011.

119 UNHCR, *Summary of Deliberations from the Expert Roundtable: Change and Displacement: Identifying Gaps and Responses*, Bellagio, Italy, 22–25 February 2011, 4.

120 "Chair Person's Summary,"*Nansen Conference on Climate Change and Displacement in the 21st Century,* Oslo 6–7 June 2011. www.regjeringen.no/globalassets/upload/ud/vedlegg/hum/nansen_summary.pdf.
121 Interview with Kenyan Commissioner for Refugee Affairs, 7 June 2011, Oslo.
122 Interview with Kenyan Commissioner for Refugee Affairs, 7 June 2011, Oslo.
123 "Statement by António Guterres, United Nations High Commissioner for Refugees, United Nations Security Council Briefing 'Maintenance of International Peace and Security: New Challenges to International Peace and Security and Conflict Prevention,'" United Nations Security Council, 23 November 2011. www.unhcr.org/4ee21edc9.html.
124 Ministerial Communiqué, Intergovernmental event at the ministerial level of Member States of the United Nations on the occasion of the 60th anniversary of the 1951 Convention relating to the Status of Refugees and the 50th anniversary of the 1961 Convention on the Reduction of Statelessness (7–8 December 2011), HCR/MINCOMMS/2011/6, 8 December 2011 and Mexico Pledges, Intergovernmental event at the ministerial level of Member States of the United Nations on the occasion of the 60th anniversary of the 1951 Convention relating to the Status of Refugees and the 50th anniversary of the 1961 Convention on the Reduction of Statelessness (7–8 December 2011). Interviews with UNHCR member state representatives c and e, 8 and 10 May 2012, Geneva.
125 Nansen Initiative, *Nansen Initiative Work Plan 2014*, www.nanseninitiative.org/nansen-initiative-background.
126 UNHCR et al., *Report from Expert Consultation on Planned Relocation, Disasters and Climate Change: Consolidating Good Practices and Preparing for the Future,* San Remo, Italy, 12–14 March 2014.
127 UNHCR, "Climate Change Drives People into Harm's Way, Says UN Refugee Chief," 21 June 2012. www.unhcr.org/print/4fe3129d6.html.
128 José Riera "Challenges Relating to Climate Change Induced Displacement". *Remarks by Mr José Riera, International Conference "Millions of People without Protection: Climate Change Induced Displacement in Developing Countries,"* Berlin, 29 January 2013.
129 UNHCR, Remarks by Volker Türk, Director of International Protection, *Discussion Forum on Climate Change*, Berlin, 17 June 2014.
130 UNHCR was granted an exclusive mandate for protection at the global and national level for conflict related IDPs but not natural disaster IDPs.
131 Agnes Hurwitz, UNHCR Research Officer, Division of International Protection, *The Nansen Principles on Climate Change and Displacement: Implications for Policy and Research*, Presentation at the Refugee Studies Center Seminar, Oxford University, 1 November 2011.
132 Bryan Deschamp, Michelle Azorbo, and Sebastian Lohse, *Earth, Wind and Fire: A Review of UNCHR's Role in Recent Natural Disasters* (Geneva, Switzerland: UNHCR, 2010).
133 Loescher, *The UNHCR and World Politics: A Perilous Path*. 489.
134 UNHCR, *UNHCR's Role in Support of an Enhanced Humanitarian Response for the Protection of Persons Affected by Natural Disasters* (Geneva, Switzerland: UNHCR, 2011).
135 Betts, Loescher, and Milner, *UNHCR: The Politics and Practice of Refugee Protection*. In addition, on some occasions refugees were also

caught up in the floods. This was the case in the Pakistan floods and also in Kenya where the Dadaab refugee camp has been flooded several times in the past five years. Interview with UNHCR Senior Official, 11 November 2011, Sydney.

136 Ibid., 148.

137 Anonymous, "The UNHCR Note on International Protection You Won't See."

138 Interview with former UNHCR staff member, 27 January 2012, Oxford.

139 Interview with former UNHCR staff member, 27 January 2012, Oxford.

140 Interview with former UNHCR senior official, 27 January 2012, Oxford

141 Betts, Loescher, and Milner, *UNHCR: The Politics and Practice of Refugee Protection*, 76.

142 Telephone interview with UNHCR staff, 13 April 2010.

143 Betts, Loescher, and Milner, *UNHCR: The Politics and Practice of Refugee Protection*, 78–79.

144 Ibid.

145 Deschamp, Azorbo, and Lohse, "Earth, Wind and Fire: A Review of UNCHR's Role in Recent Natural Disasters," 2.

146 Ibid., 3.

147 Ibid., 8.

148 Ibid., 9.

149 Ibid., 9.

150 Most states had major reservations about UNHCR taking on an expanded role. They were not "convinced by the information provided to date" on the need for UNHCR to take a lead in this area. The final decision was that UNHCR needed to conduct further consultations and outline in a paper to states why the IASC was proposing such a role for UNHCR. UNHCR, *Draft Report of the 50th meeting of the Standing Committee* (1–3 March 2011) (Geneva, Switzerland: UNHCR), 5.

151 José Riera, "Challenges relating to climate change induced displacement." Remarks by Mr José Riera, International Conference "Millions of People without Protection: Climate Change Induced Displacement in Developing Countries," Berlin, 29 January 2013.

152 José Riera, "Challenges relating to climate change induced displacement." Remarks by Mr José Riera, International Conference "Millions of People without Protection: Climate Change Induced Displacement in Developing Countries," Berlin, 29 January 2013.

153 The only slight exception is UNHCR Papua New Guinea's support to the government in relocation of people from the Carteret Islands. However this is not an issue of statelessness, as the population will remain within Papua New Guinea and secondly it is contested as to whether climate change is causing the islands to "sink" or if it is tectonic plate subversion. Telephone interview with Mary-Anne Loughry, Associate Director of the Jesuit Refugee Service Australia, 14 February 2011.

3 IOM and climate change

- **IOM's evolving mandate**
- **IOM and climate change (2000–2015)**
- **Conclusion**

IOM is responsible for promoting orderly and humane migration. Interestingly, it has not received the same academic attention as UN entities, such as UNHCR, the World Health Organization (WHO), or the World Trade Organization (WTO), or international financial institutions such as the World Bank and the International Monetary Fund (IMF). It is neither perceived to be a powerful institution nor a moral leader in global crises. We do not look to IOM to determine international migration policy: this issue has remained firmly in the purview of states. Nevertheless the lack of scholarship on IOM remains a puzzle given the institution's past and current role in migration and humanitarian issues. This chapter addresses an important empirical gap by examining if and how IOM has engaged with the environment and climate change.

The chapter argues that IOM staff lobbied states to make climate change a priority as they saw a clear issue-linkage between migration and climate change. Member states were initially reluctant, yet IOM found ways to work on climate change migration, until the IOM governing council (comprising 162 member states in November 2015) endorsed this work. The first section examines IOM's origins and evolving mandate. It traces IOM's engagement with the environment and migration in the 1990s and emphasizes that states did not see environmental migration as part of IOM's core mandate. The substantive section then traces four dimensions of organizational engagement with climate change: policy, structural, rhetorical, and operational changes between 2000 and 2015. In particular it highlights how IOM lobbied states to expand into climate change.

IOM's evolving mandate

Original mandate

IOM was created in 1951 as the Provisional Intergovernmental Committee for the Movement of Migrants from Europe (PICMME). Its purpose was to organize the relocation of thousands of labor migrants from postwar Europe to the Americas, Australia, and New Zealand.[1] Its primary function was "to make arrangements for the transport of migrants, for whom existing facilities are inadequate, or who could not otherwise be moved, from European countries having surplus populations to countries overseas."[2] To perform these tasks PICMME inherited a fleet of ships and "trained officers and its experience in the mass movement of manpower across oceans" from the International Refugee Organization.[3] With these assets the organization was able to transport thousands of labor migrants and refugees from Europe across the Atlantic and Pacific oceans.[4]

The United States played a critical role in establishing the office. The United States was concerned by high unemployment rates in Europe, which it feared would lead to instability and support for Communism. They saw labor migration as a solution to this problem and convened an international conference in Brussels that established PICMME. Subsequently, PICCME became a permanent organization, losing its provisional status, and was renamed in 1951 the Intergovernmental Committee for European Migration (ICEM). It was established outside of the emerging UN structure as the United States refused to fund an international migration organization with Communist members.[5]

Mandate expansion

From the 1960s, ICEM began "migration as development" programs where it recruited highly skilled migrants from Europe and placed them in Latin America.[6] Then in 1980 the agency went global, dropping European from its name and becoming the Intergovernmental Committee for Migration (ICM). It replicated its "migration for development" programs elsewhere in Asia and Africa and a number of new non-European members joined including the Philippines, Thailand, the Republic of Korea, and Kenya.

The end of the Cold War in 1989 bought a major change for the organization and took on a new constitution and a new name—the International Organization for Migration. These name changes and new constitution reflected IOM's global ambit and broader scope. The most significant changes in the 1989 constitution were the deletion of its focus on European migration; a new emphasis on a broader range of people requiring assistance; and the addition of new functions to its purpose. These functions included the provision of "migration services" such as recruitment, language training, medical examination and reception, integration activities, and research on international migration.[7] However, unlike UNHCR, IOM does not have a mandate for the development and setting of norms nor does it have a legal protection mandate.[8] One member state explained that IOM is "much more like a service provider. It has a constitution but not a convention [such as the Refugee Convention] but the constitution is just a founding document."[9] The agency requires a request from a member state or from the UN to carry out activities.[10]

Another significant difference is that IOM receives the majority of its funding through earmarked projects. In 2008 over 96 percent of its funding was earmarked, leaving less than 4 percent of its budget from regular contributions of member states.[11] As IOM does not receive a large regular budget of non-earmarked funds it has less autonomy to choose which areas to work in, it rather carries out activities which are funded by a donor. The United Kingdom's Department for International Development (DFID) noted that: "IOM has a market-oriented approach as a reactive project based organization offering migration services in 12 broad areas of activities but is limited in its ability to direct resources strategically."[12] IOM displays a trend of "bilateralization" seen in many UN programs and agencies.[13] States influence the organization's policies through bilateral financing, rather than decisions taken at the executive board or council.

IOM also hires staff on a project basis. As a result staff are at risk of losing their jobs when the project ends and it is "difficult to create any new positions [in IOM] without a project."[14] Ninety-seven percent of IOM's staff are in the field implementing projects, which leaves a small staff of three percent at headquarters working in strategic, administrative, and oversight roles as documented in the organigram in Figure 3.1.[15] IOM is dependent on the acquisition of new projects to maintain the jobs of its employees. One scholar even compares it to a company that produces only those goods that have been ordered in advance.[16]

IOM donors are more likely to target their influence through their funding decisions for projects rather than by lobbying for changes in policy at headquarters. The United Kingdom for instance argued that IOM "is an independent project-based organization which DFID and UKBA [UK Border Authority] only seek to influence on a project

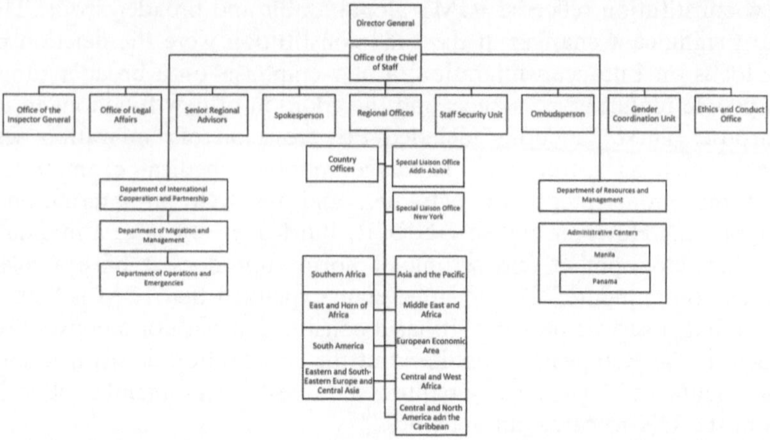

Figure 3.1 IOM organigram (2015)
Adapted from: *IOM Organizational Structure*, www.iom.int/sites/default/files/Organigram.pdf.

basis."[17] Most states spend less time monitoring IOM at headquarters than they do for UNHCR, and many states even monitor IOM from their capital.[18] Furthermore, some states still perceive IOM as predominantly a "travel agency" responsible for migration services and thus the lead ministry working with IOM is the ministry of immigration or justice, rather than foreign affairs.[19] States are also less concerned with policy or mandate matters at IOM council meetings.[20] In fact, several states claimed that "member states don't talk about mandate" and that IOM is "more interested in filling a gap if they can find funding for it."[21] Thus IOM has a high degree of operational autonomy: states may choose not to fund IOM's expansion into a new area but they are also unlikely to strongly oppose it if it finds funding elsewhere.

IOM and the environment (1990–1999)

PICMME and ICEM were not established with a mandate for climate change or environmental migration. However, in the 1990s, IOM expanded into humanitarian assistance, in particular in natural disasters and environmental migration. This expansion was contested by member states but provided a foundation for their later engagement with climate change, as this section examines.

IOM's new (1989) constitution mandated IOM to work with refugees (in coordination with UNHCR); voluntary and/or economic migrants; IDPs and other "individuals in need of international assistance."[22] This last category was described as the "grey zone" migrants, and included those that did not easily fit into the traditional categories, such as stranded migrants, IDPs and those displaced by natural disasters or potentially even climate change—although the latter was not seen as a cause of displacement at the time.[23] This new broader mandate enabled IOM to move into humanitarian work, working with those displaced by conflicts and natural disasters in the 1990s and 2000s. However, the ambiguity of the term "individuals in need of international assistance" engendered debate between IOM and member states over the boundaries of its mandate.

IOM expanded its operations into humanitarian work in the 1990s as part of a global paradigm shift in the post–Cold War era. As the Iron Curtain fell, the distinction between refugees and migrants was no longer as clear cut.[24] According to the IOM legal advisor Richard Perruchoud, there were more diverse and complex causes interacting to create population movements.[25] There was also increased funding from the international community for humanitarian operations and in 1992 IOM established its first Emergency Relief Unit. Its goal was to "allow IOM to react more promptly and in a more systematic way to crises."[26]

However, IOM's emergency work was not supported by all member states—concerns were raised during meetings over the IOM strategic plan in 1995. Member states took issue with IOM's desire to provide assistance to people affected by emergencies. They did not want the agency to overlap with other humanitarian agencies and recommended that it should consider any activities in this area as additional and not core priorities. Nevertheless, IOM became increasingly involved in relief efforts in conflict, political crises, and natural disasters. In 1998 it conducted its first operation in a natural disaster zone in Honduras after Hurricane Mitch struck.[27] The Emergency and Post-Conflict Relief Unit was restructured in the late 1990s and then reestablished in the early 2000s, and its work continued to expand.

In addition to humanitarian activities, IOM developed specific expertise on the environment and migration in the 1990s. It organized two conferences and several research reports on the issue. In 1992 it convened its first conference on migration and the environment. This was spurred by the 1992 UN Conference on Environment and Development at Rio de Janeiro, which placed the environment on the global agenda. IOM's 1992 conference and subsequent work focused on environmental change as a cause of migration and on migrants' impact on the environment. The director general, James Purcell, outlined that the conference was convened to "address the causes of environmental migration, understand the needs of those forced from their homes by environmental degradation, … and develop … more effective strategies to minimize negative impacts on the environment."[28] There was no focus on climate change, as distinct to environmental migration, at this conference, despite the fact that the first report of the IPCC in 1990 stated that "the gravest effects of climate change may be those on human migration."[29] The 1992 conference and subsequent report set IOM's priorities on "environmental migration" for the years to come and was followed by two more conferences in 1996 on the topic.

By the late 1990s IOM had expanded its mandate to encompass a wide range of migrants, IDPs, refugees, and other displaced peoples. It was an active humanitarian agency and had developed specific research expertise on environment and migration. IOM had framed a new policy problem of "ecological migration," which it defined as "migration caused by processes of environmental degradation including worsening quality and accessibility of natural resources."[30] It sought to end the vicious circle of mutually reinforcing environmental damage and mass migration and had showcased its research and policy work at various conferences, which were supported by some member states. However, IOM had not explicitly engaged with climate change, as distinct from the environment, as a driver of migration.

Does this mean that IOM had a mandate to work on climate change induced displacement and migration? One senior IOM official claimed that climate change and migration "was not a mandate issue for IOM" because IOM has the authorization to work on it.[31] Yet, this same senior IOM official also acknowledged that many developed countries in 2006 claimed climate migration was, "not part of the mandate." She explained that "Switzerland argued that the time was not yet ripe," and even if it was ripe, "they were not sure IOM should be working on it [climate change and migration]."[32] This view has been confirmed by other member states, one of whom expressed that "there is a view amongst member states that climate change is not an issue that … IOM should be working on."[33] In sum, the ambiguity in IOM's mandate over who constituted an individual "in need of international migration services" resulted in a gray zone of delegation. It was not clear who was under IOM's mandate and who was not.

Nevertheless states did not formally instruct IOM at the council meeting to work on environmental migration. They were reluctant for an agency to expand significantly into new areas—be it climate change, environmental migration, or humanitarian activities—and some were unaware that IOM even worked in this area.[34] The next section will examine how IOM negotiated this ambiguous gray zone of its mandate through its response to climate change.

IOM and climate change (2000–2015)

Between 2000 and 2006 the environment slipped under IOM's radar and there was no significant rhetoric, policy, or operational activities that tackled this issue.[35] There was only one mention of the environment, for instance, in any of the director general's annual reports to the IOM executive committee during these years.[36]A senior IOM staff member suggested, for example, that the "hiatus" was partly due to a lack of "media attention."[37] Although climate change was recognized as an important global problem in the early 1990s, it was not until 2006 that climate change became one of the top international priorities for heads of state and in the media. However in the early 2000s IOM, like UNHCR, became more engaged in natural disaster and humanitarian operations. This was not conceived of as responding to climate change displacement, but like UNHCR's work, it did mean assisting people affected by extreme weather, floods, droughts, and other natural disasters.

Natural disasters and humanitarian operations

In the 2000s IOM expanded its humanitarian activities. There was an increasing need for humanitarian assistance following natural disasters and IOM sent teams to Gujarat post-earthquake (2001); Sri Lanka post-tsunami (2004); Haiti post-earthquake (2010); and Pakistan after the floods (2010). In addition, IOM's role in natural disasters was strengthened in the humanitarian reforms process. The IASC appointed IOM as cluster lead for camp coordination and camp management in conflicts and natural disasters under the new coordination system. Dealing with natural disasters was a significant share of these humanitarian activities. Between 2005 and 2008 IOM received US$300 million for natural disasters and projects relating to fast and slow onset climate change, equating to approximately 25 percent of the funding received by the department.

IOM gained a measure of autonomy from being the camp coordination cluster lead as it acquired legitimacy vis-à-vis the UN system. In 2006 the director general explained IOM's new cluster lead role to states at the council, as they had not given it a mandate to take on this work. He stated that:

> IOM was now a major disaster relief agency, and the IASC had recognized its role in the new cluster process and given it special standing with regard to natural disasters, i.e. emergencies that were not caused by war, oppression or human rights violations. Such disasters seemed to be increasing in number and duration, prompting IOM to focus more attention on them.[38]

There was no officially recorded response at this meeting to this new stream of work from member states.

States supported this work tacitly even if they did not financially. One member state representative, for instance, explained that "IOM does a lot of important work that you don't find in their mandate."[39] Another member state explained that "at Geneva we see them as a migration agency" but argued that they (IOM) "don't have to prove [humanitarian operations] is part of their formal mandate" as long as "they prove operationally sound."[40] Thus in the 2000s IOM was able to expand into assistance to those affected by natural disasters. However, IOM's natural disaster activities were not explicitly framed as "responding to climate change" and I will now turn to activities that were.

Attempted mandate change

In 2006 the organization appealed to states to fund a small meeting of academics, policy makers, and experts on environmental migration. However, states were not supportive of IOM working on the issue and did not fund the conference. Instead, IOM turned to the United Nations Population Fund (UNFPA) for funding and co-organized a seminar in February 2007 in Bangkok on environment and migration.[41] The meeting was held in the same month as the release of the fourth IPCC report, which provided strong scientific evidence for climate change to policy makers. The IPCC report explicitly mentioned that climate change was likely to cause migration, making the issue "very hard to deny" in the words of one IOM staffer.[42] This staff member maintained that the IPCC report gave IOM the legitimacy and inspired its "willingness" to work on the issue.[43]

The Bangkok meeting was not framed in terms of climate change induced migration but in terms of environment and migration. The conference discussion adopted the same focus, framing and form as IOM discussions in 1996. Conference attendees (25 policy makers, practitioners, and researchers) discussed the definition and typologies of environmental migration—from gradual environmental change to extreme events. IOM outlined a working definition of environmental migrants as "persons or groups of persons who, for compelling reasons of sudden or progressive changes in the environment that adversely affect their lives or living conditions, are obliged to leave their habitual homes, or choose to do so, either temporarily or permanently, and who move either within their country or abroad."[44] This working definition became IOM's official definition in subsequent work. This meeting constituted a reengagement with an old policy area, rather than a significant phase of new policy development.

After the Bangkok conference, IOM brought the issue of environmental migration to the attention of its membership at the annual council meeting of 127 member states. Yet again, states were reluctant to support this, and when given the choice they did not prioritize it as a topic for discussion.[45] The main reason, according to one IOM staff member, was that states had neither awareness nor interest in the issue.[46] Another IOM staff member explained that states asked "what does IOM have to do with it? Is this [environmental migration] a real issue?"[47] IOM needed to do more research and awareness-raising to make it a priority for states.[48] IOM staff were also aware of constraints, as an IOM staff member working on the issue explained: states would be "ready when they're ready."[49] What is interesting is that IOM went directly to the council for a mandate to engage with environmental migration.

In 2007 IOM convinced states to hold a three-hour discussion on migration, the environment, and climate change at the council. The agency prepared a discussion note for this meeting and set out how environmental migration was a policy problem and solution. It was a problem for those who were displaced and could have adverse effects on the recipient country or area. The paper outlined that "gradual and sudden environmental changes are resulting in substantial human movement and displacement. The scale of the flows, both internal and cross-border, is expected to rise and have an unprecedented impact on lives and livelihoods."[50] Migration was also presented as a coping strategy and thus (provisional) solution to severe environmental change. IOM maintained that: "Increased migration can contribute to further environmental degradation, but it can also be a coping mechanism and survival strategy for those who move."[51] In the paper IOM recommended that countries of origin encourage "host states to admit environmental migrants, whether as part of labor migration schemes, resettlement programs, or humanitarian assistance initiatives."[52] This discussion note set out IOM's policy position: environmental migration should be facilitated within the available legal migration channels. It also outlined a role for IOM's role in enabling "more informed action and multi-stakeholder cooperation."[53]

The subsequent discussion during the 2007 council meeting focused predominantly on the issue of environmental migration, rather than IOM's role in addressing it. A panel of speakers, including representatives from China, Bangladesh, Greece, Cameroon, and Colombia, spoke about if and how environmental migration was a problem in their country. Greece and China advocated for developed states to support developing countries' response and disaster risk reduction. Greece even pledged it would create "special funds in cooperation with regional organizations to finance adaptation projects in Africa and small-island developing states, and cooperate with IOM on various projects."[54] Greece's remarkable statement of support for IOM in this area was because this representative was simultaneously chairing the international Human Security Network and the Organization for the Security and Co-operation in Europe and focusing on the human security impacts of climate change in these roles.[55] However, Greece's pledge of financial support was an anomaly—there was no other financial support offered for IOM to pursue this work.[56]

At council IOM did not set out a new policy position or role for the agency in environmental migration. The agency did outline a series of principles states should follow to address environment and migration. These included: effective environmental migration management; proactive policy and early action; coherent policies; and bilateral, regional, and multi-stakeholder cooperation.[57] There is no official record of states disagreeing or agreeing with these principles. IOM did not explicitly establish its role in implementing or facilitating these principles.[58] This evidence suggests there was a tacit acknowledgement of the issue but not explicit support for IOM's engagement with environmental migration. Notably throughout this period IOM framed the issue as "environmental migration" and not as "climate migration."

In 2008 IOM received the first explicit financial support to work on climate change and migration from a member state. Greece financed and cohosted a half-day long conference on Climate Change, Environmental Degradation and Migration: Addressing Vulnerabilities and Harnessing Opportunities. However Greek support was limited to 2008—the year that Greece was chairing the Human Security Network and was largely due to the leadership of Theodoros Skylakakis.[59] The conference's primary objective was to raise awareness of the human security challenge of climate change for the most vulnerable people.[60] The Director General of IOM, Brunson McKinley, spoke at the conference alongside Skylakakis and highlighted IOM's expertise and experience on climate change, environmental degradation, and migration. IOM focused the conference on the human security dimensions of climate change mobility to counter the growing securitization of climate change and perception of migration as a threat.[61] The main achievement of the conference was to bring together over 180 people from 67 countries and 33 intergovernmental organizations. This example also demonstrates how IOM could harness state support to advance its agenda.

Overall, during this period IOM lobbied member states to recognize environmental migration as critical policy issue that the organization, and states, should focus on. They did this by initiating conferences, and setting the agenda of council meetings. However they did not gain a formalized mandate change as states did not agree that it was a priority.

Structural change

In 2007 and 2008 IOM made structural changes to reflect its reengagement with environmentally induced migration. IOM established a "loose focal point" for environmental migration within the Migration Policy, Research and Communications Division to be assisted by two migration policy officers. The focal point was appointed to ensure there was a "party line" in IOM on climate change and migration.[62] In addition, there were about ten other staff spending time on climate change across IOM in research, as well as operational staff working on climate change, environment, and natural disasters. During the 2007–2008 period the climate change work was "pretty intense."[63] The focal point "pushed" the issue internally, highlighting to others that "this [issue] matters."[64] They participated in a number of events with governments and universities, outlining IOM's position and sharing their technical expertise on the issue.[65] They emphasized that governments should improve their coordination between ministries—immigration, labor, and the environment. They encouraged the IOM leadership to speak on it at the UNFCCC and other forums and the focal point structure remained throughout the period, even during the 2009 period of major organizational reform. Thus structural change enabled subsequent policy and rhetoric change.

Policy change

From 2008 onwards IOM sought to establish itself as the expert on the issue of climate change induced migration. They published research reports, spoke on panel discussions, and organized awareness-raising seminars.[66] These seminars and discussions were opportunities for IOM to convince states and other agencies (NGOs, international organizations, civil society) that climate change and migration was an important issue within IOM's expertise. These were not significant policy developments but demonstrated IOM's growing expertise. Although most of the reports claimed that establishing a causal connection between climate change and migration was highly problematic, they still advocated for further investigation of the link and developed methodologies and typologies to pursue this research goal. By late 2008 there was a widespread view within IOM policy and research units that climate change induced migration was a phenomenon that was worthy of further research and an important issue which IOM should pursue. Yet staff were simultaneously aware that there was not a simple causal connection between climate change and migration.

IOM also took initiatives to establish collaborations on climate change migration with other organizations. As mentioned in the previous chapter, in 2008 IOM instigated a working group on climate change, displacement, and migration under the IASC. This was an important initiative as it enabled a space away from member states to discuss climate change migration, develop common humanitarian policies across agencies, and write submissions to the UNFCCC. In October 2008 the working group submitted their first working paper, "Climate Change, Migration and Displacement: Who Will Be Affected?" The paper outlined IASC's commitment to "take account of, and manage, the humanitarian consequences of climate change, including protecting those who may move as a result" and to "launch a dialogue among Member States on how to fill existing and foreseeable legal, operational and capacity gaps associated with climate change and human mobility."[67] In 2009 they made three more submissions to the UNFCCC and lobbied for the final agreement to include migration as a possible consequence of, and adaptation strategy to, climate change.[68] The IASC working group shaped IOM's work significantly. It gave them a structured space to interact with other agencies and to develop policy and advocacy frameworks. In addition, the rhythm of the UNFCCC annual meetings focused their advocacy efforts. Agencies had to agree on submissions for each UNFCCC meeting and the IASC was a catalyst for developing IOM's policy on climate migration.

IOM also established a new Climate Change, Environment and Migration Alliance (CCEMA) with UNEP, the United Nations University (UNU), Munich Re Foundation, and civil society partners. This alliance's primary purpose was to develop policy approaches and research to investigate the links between climate change, environmental degradation, and migration. It aimed to support the most vulnerable countries with capacity-building, incentivize support for sustainable livelihoods projects with migrants, and work with national governments to combat the degradation of natural resources. It was a broad and ambitious agenda ambit for a small alliance. IOM's participation in this alliance signaled its ongoing interest and commitment to the issue and provided a forum to engage away from member state monitoring.[69]

In May 2009, with Copenhagen in sight, IOM published its first policy paper explicitly on climate change and migration. The nine-page brief "Migration, Climate Change and the Environment" outlined the "complex" relationship between climate change and migration.[70] It stated the "irrefutable evidence regarding climate change" and expectation that global migration flows would "rise significantly over the next decades as a result of climate change."[71] The paper emphasized that the agency had a "long established" interest and expertise in the area through its publications and research and operational responses to natural disasters.[72] It outlined ambitious future goals to mainstream climate change and environment into migration policies, and to minimize forced displacement by "developing temporary and circular labor migration schemes with 'environmentally-vulnerable' countries."[73] The policy statement was a signal that IOM had the necessary expertise, experience, and strategic direction to address climate migration.

In summary, IOM shifted from a focus on environmental migration (1990s) to a focus on climate change and migration (2008 onwards). It also called for new policy responses: the creation of international legal frameworks for those outside the refugee and IDP frameworks, the development of new migration policies to facilitate movement and the integration of migration as adaptation into the UNFCCC text through IASC submissions. These policy changes were significant given IOM has a small headquarters with little policy-making capacity.

Rhetorical change

IOM had two director generals, both former American ambassadors, during this period: Brunson McKinley (1998–2008) and William Lacy Swing (from 2008). Neither were leading champions of expansion into climate change migration—unlike Guterres they did not instigate their organization's engagement with the issue. However they both supported, publicized, and advocated for IOM's work in this area. For instance, in March 2008 McKinley explicitly stated that "The International Organization for Migration has an obvious role in addressing the linkages between environmental degradation, climate change and migration."[74] However, he did not attend the UNFCCC summit in 2008 but was represented by the climate focal point, Philippe Boncour.

Boncour argued for two paradigm shifts: "1) from migration as a worst-case scenario to migration as an adaptation strategy, and 2) from reactive to proactive thinking: from an ex post response to ex ante prevention and preparedness."[75] At Poznan IOM's principal interest was to ensure migration was seen as an adaptation strategy within the UNFCCC text and not as a "failure of adaptation."[76] This position was supported by other agencies in the IASC and CCEMA. IOM's engagement with the UNFCCC marked a shift in its policy and advocacy focus.

Reinforcing this role, McKinley announced to states at the annual IOM council meeting in November 2008 that climate change was an area of strategic priority. There is no officially recorded response from states on this. However, member states at this meeting expressed concern about mandate creep. The organization was urged to consolidate its work in line with the 12 strategic activities. Particular disquiet was expressed about the possibility that IOM would stray from helping member states formulate migration policy and take on a normative role.[77]

The director general responded that there "should be no mandate creep" and pledged that IOM would always provide compelling evidence of linkages between its work and the 12 strategic activities established in 2007.[78] In addition, he stated that one of IOM's five "broad strategic directions" was to "engage cooperatively and thoughtfully in emerging fields such as elections and climate change."[79] IOM could claim it had tacit consent, based on no vocal disagreement, for continuing research, conferences, and submissions on climate migration.[80] It is likely that the agency would have faced strong opposition from states if it had sought a protection role for climate change displacement.

In December 2008, William Lacy Swing, with support from the United States, replaced the incumbent American Brunson McKinley who was controversially seeking a third term in office.[81] Swing was reportedly reluctant to make climate change induced migration a priority. According to one senior IOM official he was "reluctant because of the evidence. [We have] seen such widely different statistics, facts, and a range between them is so huge that [he] needed to be convinced that it is climate change as opposed to some other factors causing the displacement."[82] Thus IOM staff had to do a "lot of ground work" to convince him to engage with the issue and then later to attend Copenhagen.[83] They "spent a lot of time reviewing the key evidence" and argued that while climate change was not the "decisive factor" there was no question that it was a "contributing factor."[84]

In the lead-up to Copenhagen, Swing did speak out frequently on climate change and migration. In September 2009 he made a speech at the UN in New York on emerging policy perspectives on human mobility in a changing climate. He highlighted IOM's contribution in carrying out "relevant operations" in over 40 countries, developing a research base, setting out the policy issues, and working in partnership with other agencies. He also emphasized IOM's goal of "raising the awareness of human mobility" within the UNFCCC process. Then in December Swing published an Op-Ed in the French newspaper, *Le Monde*, where he called on the international community to accept the principle of greater mobility of people who must migrate, temporarily or permanently, in order to adapt or to survive climate change.[85] This Op-Ed was notable as it was similar in style, timing, and purpose to Guterres' Op-Ed. The core message was that climate change induced migration was a problem that the UNFCCC and the international community needed to address.

In December 2009 Swing was the first IOM director general to attend the UNFCCC. He spoke at a side event, Climate Adaptation Continuum, Migration and Displacement—Copenhagen and Beyond, alongside the High Commissioner for Refugees and other IASC working group members. At this event he emphasized IOM's expertise in working with "environmentally displaced persons." He reiterated the long institutional engagement of IOM with these migrants:

> Certainly since Hurricane Mitch in Central America in 1998, IOM, together with its humanitarian partners, has been there every time a major disaster struck and forced populations to flee for sheer survival. We know how to put up the tents in displacement camps, we know of the protection and assistance needs of displaced persons, we know how important it is to build back better.[86]

He argued that migration should not be a strategy of "last resort" but that the international community needed to respond sooner and see migration as an adaptation strategy.

Throughout 2010, IOM continued to speak on climate change at a range of events. They spoke at a seminar on the International Security Implications of Climate Change, at the Association of Southeast Asian Nations (ASEAN) Regional Forum (November 2010), and at the African Development Forum (October 2010). They published various reports and held a conference in Bangladesh on climate change and migration. IOM was also applauded at the Global Forum on

Migration and Development in Mexico as "perhaps the most important international organization in this area [of environmental and climate migration]."[87] Swing again attended the 2010 UNFCCC summit and emphasized that "today's reality is that climate change and environmental degradation are already triggering migration and displacement. In the past decade alone, for example, IOM undertook some 500 projects for a total of US$280 million to assist victims of environmental degradation."[88] He reiterated that migration was not a "worst case scenario" but that it "should be part of our response to climate change."[89]

Operational change

In 2009, for the UNFCCC summit in Copenhagen, IOM compiled a compendium of its climate change and environmental migration operations.[90] The Geneva headquarters invited 40 missions to send in descriptions of projects that related in some way to climate change and the environment.[91] The resulting *Compendium of IOM's Activities in Migration, Climate Change and the Environment* covered a broad range of activities in 30 countries—from Angola to Azerbaijan, from Timor-Leste to Trinidad and Tobago. The compendium was a major enterprise due to the decentralized nature of the organization, and the process of producing it elevated the issue internally.[92] In the process of compiling the report IOM staff in headquarters and in the field became aware that a lot of work "has already been done on climate change and environment."[93] The 300-page compendium was a useful external document as it illustrated that IOM was already an operational expert in the area of environment, climate change, and migration. In fact, it was so popular with participants at the Copenhagen summit that IOM ran out of copies to distribute.[94]

However, the compendium also raised important questions on what constituted an environmental or climate change project. Entries included activities as varied as providing earthquake shelter assistance in Pakistan, soil conservation and reforestation in Haiti, and promoting youth employment in the environmental sector in Senegal. There was no clear, common skill that IOM was delivering across these projects and as these examples illustrate, some of the listed operations had no explicit link to climate change adaptation and/or only a tenuous link to IOM's migration mandate.[95]

Furthermore, the content of the activities depended on what donors were prepared to fund. One member state, for example, visited IOM's reforestation activities in Haiti. They claimed these activities were not in IOM's "core mandate" and not a "core capacity of IOM."[96] They acknowledged that "mission creep" was occurring but did not see this as a "dangerous development" as they argued "someone needs to do it [reforestation]."[97] Nevertheless this state would not fund IOM's reforestation or other natural disaster activities as it only financed core mandated operations, in particular IOM's assisted voluntary returns program.[98]

Moreover, the compendium inadvertently highlighted the disconnect between the global policy debate and operations on the ground. IOM's activities dealt with a range of migrants and nonmigrants in situations where environmental change was sometimes severe and sometimes not even apparent. People did not fit within the clear typologies of environmental migrants that IOM had developed. Moreover, climate change was not identified as either a direct or contributing cause of many of the activities. The compendium highlighted a conceptual ambiguity and tension between IOM's climate operations and policy statements, examining IOM's climate change adaptation programs in Kenya offers further insights into this.

Climate change adaptation in Kenya

In 2010 IOM initiated work with pastoralists in northern Kenya on climate change adaptation. The Livelihood Support to Pastoralist Communities and Refugees' Host Communities in Response to Climate Change and Refugee Influx in Northern Kenya was designed to assist pastoralist communities in dealing with drought, which the Kenyan president had declared a national emergency in 2009.[99] This one-year climate change adaptation project continued in 2011 as Mitigating Resource Based Conflicts among Pastoralist Local Communities including Refugee Host Communities in Northern Kenya through Strengthening Youth Capacities to Adapt to Climate Change.[100] Initiatives under this work stream included: a poultry project for five women's groups, installation of 25 sprinkler irrigation systems procured for vegetable farming, planting over 200 aloe vera seedlings, consultations with water resource authority on sinking five boreholes, provision of 15 timber boats and 2 fiberglass boats to fishermen in Lake Turkana, and the establishment of a bone craft training center in Kakuma.[101] On a given day the project coordinator dealt with issues as diverse as finding a spade for a greenhouse gardener, liaising with a mayor on where to build stalls for a market, to instructing farmers where to plant and irrigate crops.[102]

Interestingly, most of IOM's climate change adaptation activities in 2010 and 2011 were the same as earlier livelihood activities in 2009.

The operational activities stayed largely the same while the labeling of them changed. IOM staff in the field did not clearly distinguish between what was a climate change adaptation activity and what was not. In discussions with the project manager he identified a number of projects as adaptation, such as shallow water wells, planting of neem trees, drip irrigation, provision of boats to fishermen on Lake Turkana, and integrated small-scale farming.[103] Yet no IOM field staff provided a clear rationale of what distinguished these from the other activities. In fact, the IOM field staff stated that a number of activities could be shifted between climate change adaptation and other project categories.[104] This may be in part due to the ambiguity of the term adaptation—almost any development activity can be subsumed in the term under some definitions as discussed in chapter two.

Importantly, the shift from livelihoods to climate adaptation was not a result of requests from donor agencies. Japan made no explicit demands that the project deal with, or be labeled, climate change.[105] Instead, it appears that IOM staff decided to label the program as a climate change initiative. IOM consulted the Japanese donors in advance—at both the Nairobi and Tokyo level—and they supported the change in labeling of the activities. This was facilitated by the "very, very flexible" Japanese funding which was originally allocated for similar projects on "food security" in the Rift Valley, where they had a surplus.[106] In fact, the project manager had previously worked in the Rift Valley and transferred the portfolio of programs designed for the fertile Rift valley to Kakuma.[107]

IOM's livelihood and climate adaptation activities were outside the organization's core expertise and not closely linked to its migration mandate. IOM developed, for the first time, working relationships with the Food and Agriculture Organization (FAO) and the Ministry of Agriculture and the Ministry of Livestock to implement these projects as it did not have the expertise needed in livestock and agriculture. IOM made a number of claims to justify their expansion into this new area. They referred to the need to take a "holistic approach"[108] and the fact that there was an institutional gap[109] as no other agencies worked with host communities or pastoralists. Most importantly they claimed that pastoralists were migrants, and thus any work with pastoralists fitted their mandate. As one IOM staff member explained, "We are not an agricultural organization, so most of the time people would ask, why are you involved in this? Anything to do with migration is our field—pastoralism is our field."[110] Essentially, IOM staff in Kenya perceived their mandate as very broad and there were few projects that could not be linked to it.

One donor questioned this logic for expansion. He asked IOM in a bilateral meeting if working with pastoralists was "really part of your [IOM's] mandate and they said 'sure, migration is what we do.' [But] it's a more complex issue than that… . I understand that migration is part of their mandate but still I'm not quite sure if this is what they have expertise on."[111] Yet as this donor pointed out:

> These are very marginalized areas both politically and economically and … difficult areas to work in. You need a very good understanding of how the livelihood is, the coping mechanisms, the whole climate change, how have pastoralists adapted to climate change in the past twenty years. How much is a coping mechanism and how much is traditional? I think there are probably UN agencies and NGOs that have been tracking and following that conversation for a longer time than IOM.[112]

IOM's broad mandate enabled it to easily shift into a new area but it did so without the necessary expertise. They replicated their programs from the fertile Rift Valley in the arid Turkana region, and many did not succeed.[113]

There are risks involved in expanding into a new issue area without the adequate expertise or institutional support. A number of government officials criticized IOM's initiatives as not appropriate and not suitable. They referred to mistakes with IOM's selection of crops to grow (such as bananas) in arid areas.[114] IOM's projects were also described as "not well thought out" and "destined to fail" because of the way of identifying beneficiaries.[115] Moreover, IOM projects were not informed by a long-term vision of sustainable, nondependent, development in northern Kenya.[116] Most of the initiatives seemed to be small-scale, once-off injections of capital and resources, which may not lead to sustainable income generation for the beneficiaries. One government official noted that: "If you just buy items and give them to the community without thoroughly sensitizing them and letting them know why you are giving them such items sometimes it comes out as just a mere donation and you will create a dependency situation" and suggested this was what IOM was doing.[117] Another local councillor stated that "they haven't trained a person how to sustain a project."[118] They claimed they were marginalized from the projects and that there was no involvement of community leaders in project development.[119]

Overall, IOM appeared to make the largest changes at the operational level. It had no climate change and migration portfolio in 2000, and by 2010 claimed to have over 500 projects in over thirty countries worldwide.[120] IOM documented and promoted its operational expertise in the environment, climate change and migration through the 2009 compendium and at the International Dialogue on Migration in 2011. However some of IOM's new operations were a relabeling of existing projects and did not necessarily respond to climate change, as the Kenya example illustrates, although it should be noted that this project is not necessarily representative of all IOM's climate change initiatives. Furthermore, IOM's flexible bilateral funding structure enabled staff to label various initiatives as climate change related. Staff on the ground, at least in Kenya, did not see a need to substantiate the causal linkages between climate change and migration as IOM's mandate was perceived as sufficiently broad.

Mandate change

In 2009 and 2010 IOM gained support from member states to continue working on climate change induced migration. At the 2009 council meeting some states maintained that an area of "special importance" to IOM's was "climate change and the consequent displacement of migrants."[121] Again in 2010 member states endorsed IOM's work on climate change and migration at council. IOM elevated climate change to its 2010 strategic review arguing that as "emerging issues with implications for migration, such as climate change, continue to rise on the global agenda, it may also be in member states' strategic interest to ensure that IOM is tasked to specifically address such new challenges in the future."[122] At the council meeting member states agreed that the International Dialogue on Migration in 2011 should focus on climate change and migration. This was significant as the International Dialogue on Migration is IOM's top policy-level forum. States' engagement here indicates some support for IOM to work on climate migration.[123]

In March 2011 IOM convened the International Dialogue on Migration on Climate Change, Environmental Degradation and Migration, set the agenda, and 221 people attended, including 151 representatives of member states. The deputy-director of IOM, Laura Thompson, highlighted that in the previous 10 years, IOM had received funding for more than 500 projects to respond to environmental migration. IOM's aim was to bring the "topic to the table" and then let states decide if and how they would pursue it.[124] IOM did not advocate for a particular outcome from the conference and did not stipulate what its role was in implementing the conference recommendations.[125]

In addition, IOM sought out financing from new sources other than member states. IOM had successfully lobbied for the inclusion of migration as an adaptation strategy in the final UNFCCC agreement at Cancún.[126] This was applauded as a significant victory on the basis that IOM would have access to the adaptation fund. Subsequently IOM did a "mapping" of potential "use of the adaptation fund, the Least Developed Countries Fund, EU funds, as well as other bilateral, multilateral, and private sources."[127] IOM could not directly access the adaptation fund and so established a partnership with the ADB to access the fund.[128] The director general held bilateral meetings with the ADB to develop this partnership and also explored funding for adaptation projects with the Swedish International Development Agency. IOM was proactive in sourcing financing.

By 2013 IOM's policy agenda relating to natural disasters, climate change, and environmental migration was spread in four key policy debates.[129] Firstly, they sought to ensure migration was recognized as a driver of risk in the Hyogo Framework for Action discussions on disaster risk reduction (DRR) and resilience and contribute to the UN system-wide action plan on DRR. Secondly, they also advocated for the rights of migrants to be recognized at the UN Rio+20 conference—and got this acknowledged in the final outcome document. Thirdly, in the UNFCCC IOM lobbied for states to deliver on their promise to consider rehabilitation and compensation for migration under the "loss and damage" domain. They also advocated for states to integrate migration as a positive adaptation strategy in National Adaptation Programs of Action. Fourthly, in the humanitarian sphere, IOM collaborated with other agencies and pushed its Migration Crisis Operational Framework to look at vulnerable mobile groups and participated in the Nansen Initiative's steering committee. In fact, like UNHCR, IOM had become more engaged in natural disaster operations in the 2000s.

IOM continued to be involved in natural disasters in the 2010s in particular as co-lead of the camp management and camp coordination cluster. When Typhoon Haiyan struck the Philippines (December 2013) IOM brought in over 40 international specialists and increased national capacity to 50 staff members, and opened new suboffices in five areas to help over 5,000 families displaced in over 60 different sites. In the Philippines they also provided health services, protection for those vulnerable to counter trafficking and gender-based violence, and communications for those displaced. In its role as co-leader of camp coordination and camp management cluster, IOM wrote a "Comprehensive Guide for Planning Mass Evacuations in Natural Disasters" (2014), in collaboration with UNHCR and the International Displacement Monitoring Centre. The report was written because they had received requests from governments for assistance in this area and sought to prepare national, regional, municipal, and other levels of governments to develop and refine emergency evacuation plans. This work illustrates how IOM has taken an increasingly proactive role in responding to humanitarian disasters associated with sudden-onset climate change, at the request of governments.

Conclusion

States did not establish IOM with a mandate for climate change related migration, nor did it gain one in the intervening years (1951–2001). However IOM staff, particularly the climate focal point and subsequently the director general, lobbied states strongly at council between 2006 and 2011 to prioritize the issue. IOM organized a series of conferences, wrote policy papers, conducted research, and spoke at seminars and the UNFCCC summits on how climate change would lead to greater migration. IOM lobbied for migration to be an adaptation strategy with other IASC members and established hundreds of projects related to the environment and migration in over 30 countries worldwide. Over time, by showcasing its work and the importance of the issue, IOM convinced states at council to support this work and host an International Dialogue on Migration focused on climate change.

Staff built on previous issue-linkages between the environment and migration to substantiate their work on climate change and migration. It was not a big leap for staff to justify working on climate change after their work in the 1990s on the environment. However, various IOM reports did note that there was not a clear-cut causal, direct link between climate change and migration. This made it difficult to identify if and when migration would happen, and whether migration was a problem or the solution to climate change.

Moreover, IOM was able to pursue climate change without explicit endorsement from council in the 2000–2008 period. It did so through bilateral financing from states and the IASC task force on climate change and issuing reports and research. IOM was able to work with sympathetic member states (such as Greece) to pursue this stream of work. This was partly because states did not scrutinize IOM's mandate and activities as closely as UNHCR's and because of its largely earmarked funding. IOM was able to increase its mandate by pursuing financing from states, and then lobbying states.

What does IOM's work suggest for assistance for people affected by climate change? Firstly, IOM is addressing many needs within its humanitarian operations, especially for IDPs, and these are likely to continue. However, IOM has no mandate for protection of those displaced across international borders and is not the appropriate place to elaborate new protection norms. Yet IOM will continue with its research and reports on the issue, and continue lobbying the UNFCCC to take migration into account.

Notes

1 Marianne Ducasse-Rogier, *The International Organization for Migration, 1951–2001* (Geneva, Switzerland: IOM, 2001).
2 ICEM, *Resolution to Establish a Provisional Intergovernmental Committee for the Movement of Migrants from Europe*, 1951, Brussels.
3 Gill Loescher, "The UNHCR and World Politics: State Interests Versus Institutional Autonomy," *International Migration Review* 35, no. 1 (2001), 58.
4 They also offered financial assistance to migrants, helped governments with essential migration services, and promoted technical assistance and the exchange of information about sending and receiving countries. Edward Marks, "International Assisted Migration: ICEM Rounds out Five Years of Resettlement," *International Organization* 11, no. 3 (1957): 481–494.
5 In 1946 members of the ILO had proposed a "central coordination body" to deal with "overpopulation" in Europe within the UN. However, the United States rejected this proposal and subsequently when the US congress allocated US$10 million to encourage migration from Europe they placed the condition that none of the funds could be allocated to an international organization which had in its membership "any Communist-dominated or Communist controlled country." This ruled out the ILO, and the rest of the UN, which had universal membership. Ducasse-Rogier, *The International Organization for Migration, 1951–2001*, 17.
6 Ibid., 15.
7 IOM, *Constitution of the International Organization for Migration*, Article 1 (c, d and e).
8 IOM 2007, *IOM Strategy*, www.iom.int/jahia/webdav/shared/shared/mainsite/about_iom/docs/res1150_en.pdf.
9 Interview with member state representative to IOM, 10 May 2012, Geneva.
10 IOM, *Constitution*: 1(b).
11 Fabian Georgi, "For the Benefit of Some: The International Organization for Migration and Its Global Migration Management," in *The Politics of International Migration Management*, ed. Martin Geiger and Antoine Pecoud (Basingstoke and New York: Palgrave Macmillan, 2011), 45–72.
12 DFID, *Multilateral Aid Review: International Organisation for Migration* (London: DRID, 2011), 15.
13 Erin Graham, "Money and Multilateralism: How Funding Rules Constitute IO Governance," *International Theory* 7, no. 1 (2015): 162–194.
14 Interview with IOM staff member, 7 May 2012, Geneva.
15 "Review of the IOM Strategy " IOM (MC/INF/302), 12 October 2010, 5.

16 Georgi,"For the Benefit of Some: The International Organization for Migration and Its Global Migration Management."
17 DFID, *Multilateral Aid Review: International Organisation for Migration* (London: DRID, 2011), 178.
18 Interview with IOM and UNHCR member state representatives, 10 May 2012, Geneva.
19 Interview with IOM and UNHCR member state representative, 10 May 2012, Geneva.
20 One state described IOM Council meetings as "very easy-going" and said they mostly focus on financial issues. Interview with IOM and UNHCR member state representative d, 9 May 2012, Geneva.
21 Interview with IOM and UNHCR member state representative, 7 May 2012, Geneva.
22 IOM, 1989, *Constitution*. As UNHCR takes the lead on the first category IOM has been more focused on the second two categories and is one of the only organizations that has an explicit mandate to protect this group. Ducasse-Rogier, *The International Organization for Migration, 1951–2001*, 132.
23 Ibid., 132–133.
24 Ibid., 15.
25 Richard Perruchoud, "From the Intergovernmental Committee for European Migration to the International Organization for Migration," *International Journal of Refugee Law* 1, no. 4 (1989): 501–517.
26 Ducasse-Rogier, *The International Organization for Migration, 1951–2001*.
27 "Hurricane Mitch: IOM Reconstruction Assistance for Honduras," Relief Web, May 27, 1999, http://reliefweb.int/report/honduras/hurricane-mitch-iom-reconstruction-assistance-honduras.
28 Purcell quoted in IOM, *Migration and the Environment* (Geneva, Switzerland: International Organization for Migration, 1992).
29 IPCC, *Climate Change: The IPCC Scientific Assessment* (London: IPCC, 1990).
30 International Organization for Migration (IOM), UNHCR and Refugee Policy Group, *Symposium on Environmentally Induced Population Displacement and Environmental Impacts Resulting from Mass Migration* (Geneva, Switzerland: IOM, UNHCR, RPG, 1996).
31 Interview with IOM senior official, 12 October 2010, New York.
32 Interview with IOM senior official, 12 October 2010, New York.
33 Interview with IOM member state, 23 March 2010, Geneva.
34 Interview with IOM member states, 7, 10, and 11 May 2012, Geneva.
35 Interview with IOM senior official, 12 October 2010, New York.
36 The author looked at every council report, director general's report to the council and executive committee report for the years 2000–2010 for entries mentioning climate change or environment. The one reference to migration and the environment between 2000 and 2005 was in a research paper—*Migration, the Environment and Development, the Impact of the Tsunami on Migration in Asia*—which examined the consequences of the Indian Ocean tsunami of 2004. The report, unsurprisingly, did not focus on climate change at all as the tsunami was caused by an earthquake and not a hydro-meteorological event.
37 Interview with IOM senior official, 15 December 2009, Copenhagen.

38 "IOM Report on the Hundred and Third Session of the Executive Committee," IOM (MC/2201), 26 June 2006.
39 Interview with IOM official, 11 May 2012, Geneva.
40 Interview with IOM member state representative, 7 May 2012, Geneva.
41 Interview with IOM officials, 17 March 2010, Geneva.
42 Interview with IOM officials, 17 March 2010, Geneva.
43 Interview with IOM official, 17 March 2010, Geneva.
44 International Organization for Migration IOM, *Expert Seminar: Migration and Development* (Geneva, Switzerland: IOM, 2008b).
45 Interview with IOM senior official, 15 December 2009, Copenhagen.
46 Interview with IOM senior official, 15 December 2009, Copenhagen.
47 Interview with IOM official, 17 March 2010, Geneva.
48 Interview with IOM official, 17 March 2010, Geneva.
49 Interview with IOM official, 17 March 2010, Geneva.
50 International Organization for Migration IOM, UNHCR, and Refugee Policy Group, *Discussion Note: Migration and the Environment* (Geneva, Switzerland: IOM, 2007), 1.
51 Ibid., 1.
52 Ibid., 7.
53 Ibid., 7.
54 "Report on the 94th Session of the Council," IOM (MC/2239/Rev.1), 14 February 2008.
55 The Human Security Network is an informal group of 13 states that meet regularly at the foreign ministerial level to promote the concept of human security. Interview with IOM senior official, 15 December 2009, Copenhagen.
56 "Report on the 94th Session of the Council," IOM (MC/2239/Rev.1), 14 February 2008.
57 Ibid.
58 Ibid., 30.
59 Interview with IOM senior official, 12 October 2010, New York.
60 IOM and Greece, "Climate Change, Environmental Degradation and Migration: Addressing Vulnerabilities and Harnessing Opportunities," *Report of the Conference*, Geneva, 19 February 2008.
61 Interview with IOM official, 11 May 2012, Geneva.
62 Interview with IOM officials, 17 March 2010, Geneva.
63 Interview with IOM official, 11 May 2012, Geneva.
64 Interview with IOM senior official, 12 October 2010, New York.
65 Interview with IOM official a and senior official, 17 March 2010, Geneva, and the author was also a participant at Climate Change and Migration in the Pacific Conference, University of Victoria, Wellington, New Zealand, May 2008.
66 IOM, *Report of the Director General on the Work of the Organization for the Year 2008* (Geneva, Switzerland: IOM, 2009c). In 2008 IOM published the following: 1) *Migration and Climate Change*; 2) *Irregular Migration from West Africa to the Maghreb and the European Union: An Overview of Recent Trends*; 3) *Climate Change and Migration: Improving Methodologies to Estimate Flows*; 4) *Survey on Remittances 2008 and Environment; and 5) Migration, Environment and Development: New Directions for Research.* IOM staff wrote some of these reports and

commissioned academics to write others, such as the report on *Climate Change and Migration: Improving Methodologies to Estimate Flows.*

67 IASC 2008, *Climate Change, Migration and Displacement: Who Will Be Affected? Working Paper Submitted to the UNFCCC by the Informal Group on Migration/Displacement and Climate Change of the IASC,* http://unfccc.int/resource/docs/2008/smsn/igo/022.pdf.

68 Submissions included: IASC, *Climate Change, Migration, and Displacement: Impacts, Vulnerability, and Adaptation Options,* Geneva, February 2009; IASC, *Joint Letter of the IASC Principals to the UNFCCC Executive Secretary,* 2009; IASC, *Comments and Proposed Revisions to the Negotiating Text Prepared by the Chair of the UNFCCC Ad Hoc Working Group on Long-Term Cooperative Action,* Geneva, 2009.

69 In April 2008 they also held an expert meeting in Munich to which many CCEMA members attended, including UNU, UNEP, Munich Re Foundation and with financial support from the Rockefeller Foundation. See: Koko Warner, *Workshop Report for "Research Workshop on Migration and the Environment: Developing a Global Research Agenda,"* 18 April 2008, Munich, Germany.

70 IOM, *Migration, Climate Change and the Environment* (Geneva, Switzerland: IOM, 2009c), 1 and 5.

71 Ibid.

72 Ibid.

73 IOM, *Migration, Climate Change and the Environment* (Geneva, Switzerland: IOM, 2009c), 7.

74 Brunson McKinley, "IOM statement," Institute for Public Policy Research Conference on Climate Change and Forced Migration, London, 29 March 2008.

75 IOM, "Environment, Forced Migration and Social Vulnerability: Identifying Problems and Challenges," UNFCCC Preparatory Meeting, 9 October 2008, Bonn.

76 IOM, "Migration and Climate Change: from Emergency to Adaptation," Speech at the 14th Conference of the Parties of the UNFCCC, 8 December 2008, Poznan.

77 "Report on the Ninety-Sixth Session of the Council 2008," IOM (MC/2266/Rev.1: 6), 26 November 2009.

78 Ibid.

79 Ibid.

80 There was also a discussion over whether IOM had a protection mandate for migrants or refugees. One delegation stated that "IOM did indeed have a protection mandate stemming from the IOM strategy and constitution." They added that it was "becoming increasingly difficult to distinguish between refugees and migration in the field. A factor that could hinder the effective management of mixed migration flows because institutional mandates did not appear to be in sync with reality in the field." *Report of the Council 2008*, IOM, 30.

81 *Report on the Ninety-Fifth (Special) Session of the Council*, IOM (MC/2251), 27 June 2008.

82 Interview with IOM senior official, 12 October 2010, New York; and see William Lacy Swing, Interview on *Climate Change TV*, December 2011,

http://climatechange-tv.rtcc.org/2011/12/12/william-lacy-swing-director-general-international-organisation-of-migration/.

83 Interview with IOM senior official, 12 October 2010, New York.

84 Interview with IOM senior official, 12 October 2010, New York.

85 Author's translation from the French. See William Lacy Swing, "Aidons les pays en développement à faire face aux changements climatiques," *Le Monde*, 12 December 2009, www.lemonde.fr/idees/article/2009/12/16/aidons-les-pays-en-developpement-a-faire-face-aux-changements-climatiques-par-william-lacy-swing_1281291_3232.html.

86 William Swing, Side Event: Climate Adaptation Continuum, Migration and Displacement—Copenhagen and Beyond, at the UNFCCC Summit (COP15), Copenhagen, 16 December 2009.

87 Koko Warner and Susan Martin, *Impact of Climate Change on Migration and Development—Background Paper for Civil Society Days, Global Forum on Migration and Development* (Mexico: 2010).

88 William Swing, IOM statement at the UNFCCC COP 16 Plenary session, Cancún, Mexico, 10 December 2010.

89 Ibid.

90 IOM, *Compendium of IOM's Activities on Migration, Climate Change and the Environment* (Geneva, Switzerland: IOM, 2009).

91 Interview with IOM official, 11 May 2012, Geneva.

92 Interview with IOM official, 17 March 2010, Geneva.

93 Interview with IOM official, 25 March 2010, Geneva.

94 Interview with IOM officials, 17 March 2010, Geneva.

95 Note that adaptation is a broad category so some of these projects could fit within a broad definition of adaptation, however IOM gave neither a definition of adaptation nor made any explicit connections between these activities and climate change adaptation.

96 Interview with IOM member state representative, 10 May 2012, Geneva.

97 Interview with IOM member state representative, 10 May 2012, Geneva.

98 Interview with IOM member state representative, 10 May 2012, Geneva

99 IOM, *Project with Livelihoods Component* (Geneva, Switzerland: IOM, 2011).

100 Over the 2010–2011 period IOM established a bakery, carpentry classes, small-scale greenhouse agriculture, small-scale integrated farms, a car wash business, basket making classes, a peace center, an Internet café, a radio station, beehives, stalls for local business people, planted trees, and provided soccer balls to youth. They also dug shallow wells, provided drug centers for medicating animals and restocked pastoralists' herds. The 2010–2011 Japanese funded projects are astonishing in the range and number of activities they encompass. Ibid. Interview with IOM official, 6 April 2011, Kakuma.

101 Ibid.

102 Author's experience when spending a day in the field with IOM project coordinator.

103 Interview with IOM official, 6 April 2011, Kakuma.

104 Interview with IOM official, 6 April 2011, Kakuma.

105 Interview with IOM official, 6 April 2011, Kakuma.

106 The IOM staff member said it was "very unusual" to have such flexibility with funding. This was predominantly because it came from Japan's

supplementary budget and needed to be spent within the electoral year. Interview with IOM official, 6 April 2011, Kakuma.

107 Interview with NGO staff member working in Turkana, April 7 2011, Kakuma.

108 For instance one staff member argued that it was necessary to take a holistic approach that saw peacebuilding, livelihood support, and safe-migration as a three pronged approach or a triangle "if you remove one aspect of the three the other two will collapse." Interview with IOM official, 25 March 2011, Nairobi.

109 One Kakuma based staff member stated "most of the agencies here support the refugees so the host communities are overlooked here. But the host community are also pastoralists, so that is the reason why they wanted us to identify intervention to address their food security needs." Interview with IOM official, April 7 2011, Kakuma.

110 Interview with IOM official, 25 March 2011, Nairobi.

111 Interview with IOM member state representatives, April 6 2011, Nairobi.

112 Interview with IOM member state representatives, April 6 2011, Nairobi.

113 Interview with NGO staff member working in Turkana, 7 April 2011, Kakuma.

114 Interview with Ministry of Agriculture Crop Specialist, 13 April 2011, Kakuma

115 Interview with Town Councillor, 15 April 2011, Kakuma.

116 On previous attempts by Kenyan government and colonial government to shift pastoralists into agriculture see: David Anderson and Vigdis Broch-Due (eds), *The Poor Are Not Us: Poverty and Pastoralism in Eastern Africa* (Oxford: James Curey, 1999).

117 Interview with Ministry of Agriculture Crop Specialist, 13 April, Kakuma.

118 Interview with Town Councilor, 15 April, 2011, Kakuma.

119 Interview with Town Councilor, 15 April, 2011, Kakuma.

120 IOM, *Compendium of IOM's Activities on Migration, Climate Change and the Environment.*

121 A member of the executive committee explicitly "recognized the administration's role in raising the profile of migration-related issues in the agreement expected to be produced" at the Copenhagen UNFCCC Summit. "Report of the Council 2008," IOM, 3.

122 "Review of the IOM Strategy" IOM (MC/INF/302), 12 October 2010.

123 Interview with IOM official, 11 May 2012, Geneva.

124 Interview with IOM official, 7 May 2012, Geneva.

125 IOM, "Chair's Summary of Inter-sessional Workshop on Climate Change, Environmental Degradation and Migration," International Dialogue on Migration, Geneva, 29–30 March 2011

126 *Cancun Agreements*, 2010, Paragraph 14 (f). See also Koko Warner, *Climate Change Induced Displacement: Adaptation Policy in the Context of the UNFCCC Climate Negotiations* (Geneva, Switzerland: UNHCR—Division of International Protection, 2011).

127 Interview with IOM official, 7 May 2012, Geneva.

128 Interview with IOM staff member, 11 May 2012, Geneva.

129 IOM, *Compendium of IOM Activities in Disaster Risk Reduction and Resilience* (IOM: Geneva, 2013), 11–12.

4 UNDP and climate change

- **UNDP's evolving mandate**
- **UNDP and climate change (2000–2015)**
- **Conclusion**

UNDP is the largest development agency within the UN system.[1] It provides grants to developing countries and works on capacity, good governance, and reducing poverty. It is also the lead coordinator of UN agencies at the country level and has representation in more than 120 states. Its size, global spread, normative entrepreneurship, and development focus, make it an important agency to study.[2] UNDP is much less studied than its larger counterpart, the World Bank, which has a literature dedicated to its "greening."[3] This chapter makes an important contribution by focusing on if and how UNDP has engaged with the environment and climate change, which has not previously been examined.[4]

This chapter follows the structure of the previous two: It firstly examines UNDP's mandate evolution, and its engagement with the environment in the 1990s. It documents how UNDP developed a mandate for "sustainable human development" but not climate change adaptation. The substantive part of the chapter traces the evolution of UNDP's climate change rhetoric, policies, structure, and operations between 2000 and 2014. It demonstrates how UNDP elaborated a substantive issue-linkage between human development and climate change. It argues that successive administrators played a central role in shifting UNDP towards the environment and climate change and this expansion was enabled by an increase in climate financing opportunities.

UNDP's evolving mandate

Original mandate

UNDP was created in 1965 when states merged two UN development agencies: the Expanded Programme for Technical Assistance (EPTA)[5] and the UN Special Fund.[6] The General Assembly established UNDP "to enhance coordination of the various types of technical assistance programs within the UN system" and avoid duplication of activities. ECOSOC and the General Assembly set UNDP's mandate to foster economic development in developing countries.[7] To manage UNDP, states created a new intergovernmental body, the Governing Council. However it was not a complete merger: the two funds were kept separate, the two heads worked together, and contributions could be pledged to the two programs separately.

The procedures and principles governing the EPTA and the Special Fund continued to apply to UNDP's activities. UNDP inherited a strong network of field offices from EPTA through which it delivered development assistance.[8] UNDP's assistance, following EPTA, was made up of grants, not loans, and was thus distinct from the International Bank for Reconstruction and Development, or World Bank, as it would later be known. UNDP followed EPTA's approach asking central governments to review their needs and then developed programs in the agricultural and industrial sector.[9] UNDP also continued EPTA's practice of distributing grants to other specialized UN agencies—such as ILO, WHO, or UNICEF—alongside direct grants to states. The EPTA side of UNDP offered bilateral, needs-based assistance to developing countries and other UN agencies.

Meanwhile the Special Fund had been a much smaller grants program. Paul Hoffman, the manager of the Marshall Plan, had established the Special Fund in 1958, and the United States was its biggest donor.[10] It targeted grants to developing countries to use their natural resources and develop a skilled workforce through education and technical training.[11] UNDP also continued to finance grants through the Special Fund. Notably, UNDP and its two predecessors endorsed the exploitation of natural resources for economic development. There was no mandate in EPTA, the Special Fund, or UNDP to address conservation, pollution, or other environmental issues of the era. Essentially, UNDP had a mandate as both a development fund and program. As a fund it channeled assistance from donors to other UN specialized agencies. As a program it worked with governments in over 130 countries to build capacity and develop agricultural and industrial sectors.

In 1969, at the end of the UN Development Decade, UNDP came under review. The *Jackson Report* found that the agency was "too focused on economics" and should be orientated towards the well-being of least developed countries.[12] In particular the review criticized UNDP for lacking a "think tank to work out ideas and launch them as directives for policies."[13] The *Jackson Report* recommended that UNDP advocate for human well-being rather than focusing solely on economic growth. UNDP did not make this change in the 1970s or 1980s.[14]

Mandate expansion

By the 1990s UNDP was a development agency that did "everything" but had no core specialization or focus.[15] It had neither a clientele (the UN Entity for Gender Equality and the Empowerment of Women [UN Women] and women, UNICEF and children), nor a sectoral focus like other UN agencies (the United Nations Industrial Development Organization [UNIDO] and industrial development, WHO and health). In fact the UNDP administrator in the early 1990s, James Speth, stated that it "lacked a clear substantive profile, a focus in development policy terms and a profiled strategy."[16] UNDP had evolved from its days as a merger of two funds—it had "in-house expert capacity" and it now "consumes a lot of the money it raises itself for projects it develops and executes."[17] UNDP was jeopardized by the World Bank, which in the 1980s and 1990s broadened its activities and began to "invade" UNDP's development territory.[18] UNDP in the 1980s and early 1990s was an institution in search of a purpose.

In 1994 UNDP created the Human Development Report Office and launched the *Human Development Reports*. The office was a semi-autonomous think tank within UNDP to advocate for "human development," and fulfilled the recommendations of the *Jackson Report*. The office, under the leadership of Mahbub ul Haq, elaborated a new development paradigm focused on individual well-being rather than national economic growth.[19] It published annual *Human Development Reports* with leading academics outlining the concept of human development and ranked states annually on a human development index. The Human Development Report Office had a separate governance structure and governing board to ensure editorial independence.[20] Furthermore, the office was guaranteed regular fixed contributions to finance its work and was not dependent on UNDP's core contributions.

Human development became the overarching focus for UNDP. The concept "provided technical, political and even moral guidance" and "acted as a source of institutional identity for UNDP staff."[21] As Haq explained, the human development paradigm provided the "missing moral core in our technological advance" and claimed that "in rich nations and poor, the moral foundations of economic growth are often lacking."[22] The reports, according to one UNDP administrator, created "an extraordinary advocacy tool" whose strength derives from the way it "benchmarks progress" with the human development index.[23]

Although the Human Development Reports gave UNDP a normative purpose, the Human Development Report Office was separate from the rest of UNDP and UNDP programming, as the organigram in Figure 4.1 illustrates. This gave the office editorial independence but limited its influence over UNDP policy and programming.

Meanwhile, UNDP is closely monitored by member states. Since the mid-1990s UNDP reports three times annually to the executive board, which formulates the general policy priorities and strategy, and reviews and approves programs, activities, and the budget. The board is composed of thirty-six members: eight from Africa, seven from Asia, five from Latin America, four from Eastern Europe, and twelve from Western Europe and North America. Although developing countries have a majority on the board, decisions are made by consensus and donors have significant influence as they have to finance the agency.[24] As UNDP gives technical assistance through grants, rather than lending capital like the World Bank, it is in constant need of funding from the board. The board has the power to "veto a policy line" but does not get involved with every policy detail.[25] As a member state to UNDP explained: "All major decisions about mandate have to be taken to the executive board: [we] expect that before they enter into a new issue area that they check this with their members. But [we] don't have to legislate to the very last detail. There is space for design and exploration."[26] It is this degree of space that this chapter examines.

UNDP and the environment (1990–1999)

Up until the 1990s UNDP had no specific mandate to engage with environmental issues. In fact, in the 1980s, UNDP had programs in natural resources extraction. It supported, for example, the development of Ghana's gold extraction industry.[27] There were four important changes, elaborated in this section, which shifted UNDP's focus from natural resource extraction to sustainable development. Firstly, the *Brundtland Report* (1987) outlined the concept of sustainable development and shifted the UN's development paradigm. Secondly, UNDP helped establish a new multilateral environmental fund, the Global

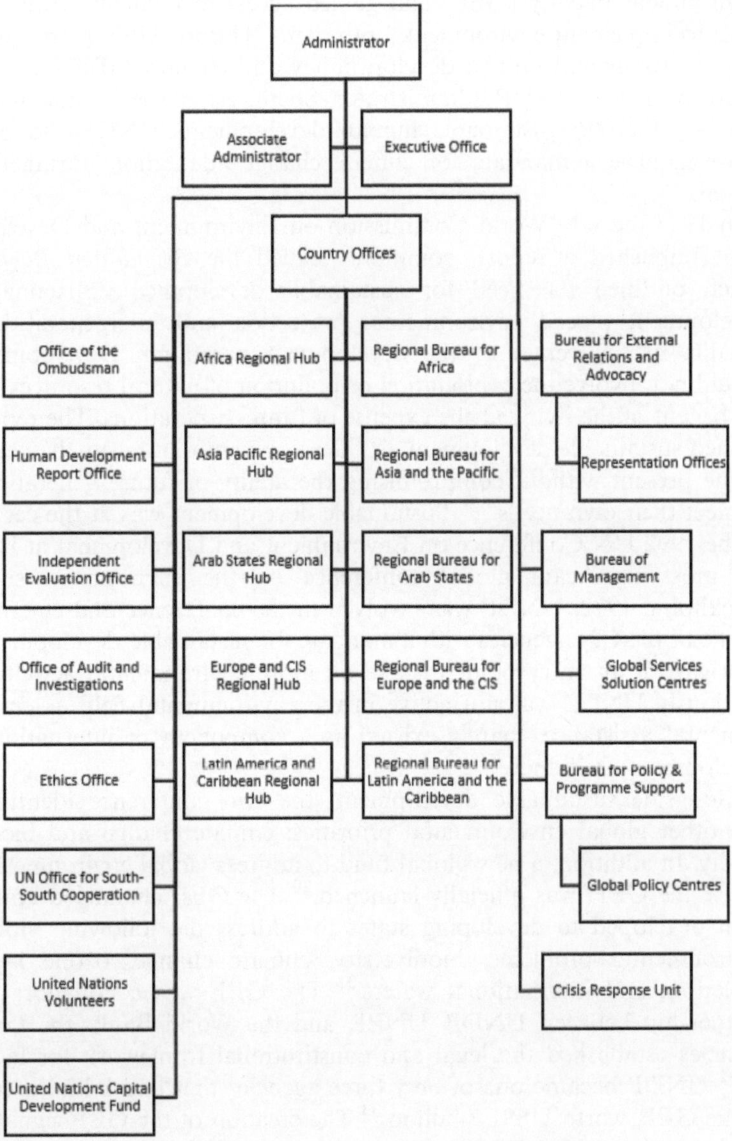

Figure 4.1 UNDP organigram (2014)
Adapted from: *UNDP 2014 Organizational Chart*, www.undp.org/content/undp/en/home/operations/about_us/organisational_chart.html.

Environment Facility (GEF) and gained access to financing from the GEF to implement environmental programs. Thirdly, UNDP created a new environmental unit to develop policy and channel GEF funding. Fourthly, a new UNDP administrator, Speth, established a new mandate for UNDP: sustainable human development. UNDP did not, however, have a mandate for climate change adaptation during this period.

In 1987 the UN World Commission on Environment and Development published a report, commonly called the *Brundtland Report*, which outlined the need for sustainable development. Sustainable development placed environmental protection and intergenerational equality at the center of the development paradigm. Development should not involve the exploitation or pollution of natural resources for the benefit of the living at the expense of future generations. The report defined sustainable development as: "development that meets the needs of the present without compromising the ability of future generations to meet their own needs."[28] Sustainable development was at the center of the 1992 UN Conference on Environment and Development at Rio, the most significant global conference on the environment since Stockholm, 1972. UNDP was involved in the conference and emerged as one of the UN agencies with a mandate for sustainable development. This gave it an entry point to take on more environmental activities; previously UNDP was not active in the environmental field as environmental assistance "barely existed as a component of international development assistance."[29]

Alongside sustainable development, the Rio conference identified two other global environmental priorities: climate change and biodiversity. In addition, a new global fund to address global environmental issues, the GEF, was officially launched.[30] The GEF channeled grants from developed to developing states to address the following global environmental problems: biodiversity, climate change, ozone layer depletion, and international waters.[31] The GEF was a collaborative partnership between UNDP, UNEP, and the World Bank: the three agencies established the legal and constitutional framework for facility.[32] UNDP became one of only three agencies that had initial access to the GEF, worth US$1.2 billion.[33] The creation of the GEF signaled the beginning of UNDP's work on climate change, in contrast to its main prior focus on energy demand, supply, and conservation.

UNDP, through the GEF, had access to a stream of financing to develop environmental activities separate from member state contributions. UNDP used the GEF to assist developing countries with "enabling activities" and implement on-the-ground activities, to fulfill requirements to the UNFCCC, the Convention on Biological Diversity, the United Nations Convention to Combat Desertification and the Stockholm Convention on Persistent Organic Pollutants from 1998 onwards.[34] Enabling activities to the UNFCCC involved monitoring, verification, and writing national communication reports, which included inventories of greenhouse gas emissions and vulnerability assessments. This was the first phase of UNDP's climate change initiatives. However, the GEF's focus on mitigation and enabling activities limited UNDP to a "very narrow understanding of what climate change meant."[35] UNDP staff did not, for instance, "understand that it [climate change] included the forestry sector, that it included all the different sectors in adaptation."[36] It was only in the 2000s that a broader conceptualization of climate change was accepted. Then "it changed fairly fast as people can see the links but they never thought of it that way until then."[37]

Meanwhile, UNDP's administrators in the 1990s sought an increasing role for UNDP in the environment. William Draper, a former Wall Street banker, and administrator from 1986 to 1993, was aware that environmental degradation was occurring and would become more of an issue in the future.[38] He also saw the environment as an area for future business for UNDP.[39] Draper established UNDP's first environmental unit: the Energy and Natural Resources Unit. In 1993 Draper was replaced by Speth, an environmentalist, who had played a central role in world environmental conferences, including Rio, and founded the World Resources Institute. Speth sought to integrate the concept of sustainable development into UNDP's human development focus. In Speth's first speech to all UNDP staff he outlined his answer: the concept of sustainable human development.[40]

Sustainable human development was Speth's way of integrating the sustainable development paradigm into UNDP's newfound focus on human development. However, he did not see anything "inevitable" about the concept, or UNDP engaging with it.[41] Speth lobbied the council to adopt a new mission statement for UNDP and claims there "was no push-back" from states.[42] In 1996 UNDP's executive board endorsed a new mission, which placed "sustainability" at its center. It stated that:

> UNDP's mission is to help countries in their efforts to achieve sustainable human development by assisting them to build their capacity to design and carry out development programmes in poverty eradication, employment creation and sustainable livelihoods, the empowerment of women and protection and

> regeneration of the environment, giving first priority to poverty eradication.[43]

This mandate change was important as it gave UNDP a more expansive role in environmental protection. However, neither was there an explicit mention of climate change in the new mandate nor any indication that this was intended at the time.

Speth also renamed the Energy and Natural Resources Unit and it became the Sustainable Energy and Environment Division (later the Environment and Energy Group).[44] The division had two functions: 1) to develop policy on sustainable development and 2) to develop GEF projects. UNDP employed a number of new staff to fulfill these goals. In 1993 UNDP headquarters hired its first climate change expert, explicitly to develop GEF climate change projects, and in 1998 it hired three more people to work on climate change as the national communications work increased.[45] However, GEF work dominated the environmental division, bringing in the majority of its funds.[46] Between 1994 and 1997 UNDP received more than US$150 million from GEF, three times the core funding of their programs.[47] One UNDP staff member claimed that the "EEG [Energy and Environment Group] was GEF."[48] GEF was reportedly such a "cash cow" that the director of the GEF unit reported directly to the administrator, instead of reporting through the director of the Sustainable Energy and Environment Division, the next-in-line manager.[49]

Furthermore, the Sustainable Energy and Environment Division's activities were siloed from other UNDP operations due to their independent funding stream. The division had little engagement or influence over the rest of UNDP's programs as it did not operate with core funding.[50] This separation was reinforced by a senior management decision to reduce funding to the Environment Division from around 20 percent to less than 15 percent of UNDP's core contributions, due to budgetary pressure.[51] This made it more dependent on multilateral trust funds. Separate GEF funding "was almost anti-mainstreaming because people had it [the environment] in a box, alone somewhere else and it wasn't important."[52] In short, UNDP's engagement with climate change in the 1990s was limited almost entirely to the GEF. As a former Environment and Energy Group staff member explained: the "GEF structured their climate change work" and "as a result UNDP never deals with climate change on its own."[53] UNDP by the late 1990s was developing and implementing a number of environmental projects, some of which focused on climate change mitigation, but these were not part of UNDP's development programming or strategic objectives.

UNDP and climate change (2000–2015)

Structural change

In 1999 Mark Malloch-Brown, a former World Bank vice president, became administrator of UNDP. UNDP was in a financial crisis: the organization's core funding was low. Malloch-Brown put it simply: "UNDP was poor and the World Bank was rich."[54] He diagnosed that UNDP had:

> attempted to compete across too many areas of development, stretching its expertise too thin. It had agriculture and education experts, public health and forestry units, urban planning expertise and much else—even though the UN had other specialized agencies in each of these areas.[55]

In doing so UNDP had built up rival expertise with other UN agencies rather than acting as the UN's development coordinator. UNDP's projects were also often tokenistic according to the new administrator and donors were demanding greater efficiency. He also argued that UNDP did not apply the World Bank's development rigor to its projects and sought to add the "professionalism" of the World Bank to UNDP.[56]

Malloch-Brown's answer was to refocus UNDP on "upstream" macro policies with an emphasis on building good governance. He argued that UNDP needed to "cut the redundancy and concentrate our resources where they made the most difference."[57] He sought to create "a highly focused"[58] organization and downgraded the environment and natural resource management "as having little to contribute to the core UNDP mandates of poverty and governance."[59] During his period as administrator, Malloch-Brown almost never mentioned climate change in speeches, indicating that climate change was not a priority during his leadership.

Malloch-Brown made major structural changes to the Environment and Energy Group. He disbanded the forestry program, reduced the number of staff working on the environment, and decentralized the Environment and Energy Group. He also discontinued positions in sustainable livelihoods, transport and sustainable development.[60] He sought to reduce the number of staff in the Bureau for Policy Development, of which the Environment and Energy Group was part, from 250 to fewer than 120 staff members at headquarters, with 98 staff redeployed to the field by 2001.[61] Malloch-Brown's decentralization and restructuring caused a sharp decline in the number of environment staff positions at headquarters and a number of the senior environment and energy staff left upon his arrival.[62]

In 2001 UNDP's environmental activities were almost exclusively GEF funded and there was no core funding for climate change activities.[63] UNDP had lost much of its environmental policy work and UNDP environmental staff would be asked at non-GEF environmental meetings, "why are you here?"[64] Yet, while the rest of UNDP was facing major cuts, the GEF unit continued to have access to substantial new financial resources. UNDP was acquiring an increased proportion of funding from the GEF, particularly during the 2003–2006 period. Staff were encouraged to develop the maximum possible number of projects likely to be approved by the GEF.[65] Thus, the environment and energy portfolio became even more dependent on the GEF. Malloch-Brown would support climate change activities as long as they were financially self-sustaining (through the GEF or other multilateral funds) and did not drain core resources. Although UNDP did develop climate change mitigation projects they were not aligned with Malloch-Brown's strategic focus on poverty reduction and good governance.[66]

Mandate change

The UNDP board was broadly supportive of Malloch-Brown's deprioritization, decentralization, and downsizing of the Environment and Energy Group. In 2000 the executive board decided to discontinue environment as a core priority within UNDP's multi-year funding frameworks.[67] Although the environment was reinstated as a priority in 2002 it, "never regained its status as a core priority supported by core funds."[68] Furthermore there was little mention of climate change in the board's annual meetings between 2000 and 2005. Climate change did not need to be a high organizational priority as it could rely on funding through the GEF during a period of declining core contributions. UNDP's overarching goal was poverty reduction and other agencies, such as UNEP, had a greater mandate and expertise in the environmental sphere.[69] Some suggest that Malloch-Brown became more supportive of UNDP's environmental activities in the 2002–2005 period.[70] This was reflected in a keynote speech he gave at the GEF assembly in 2002 for which he received a "standing ovation."[71]

In 2005 Kemal Derviş took over as Administrator. He was a former Turkish minister of economic affairs and had worked at the World Bank as the chief economist. He was a "very intellectual and solid economist and always demanded that policy decisions be rigorously backed by empirical evidence whenever possible."[72] This may have influenced his outlook on climate change, particularly at a time when many governments were reluctant to admit it was anthropogenic and the development-adaptation issue linkage was weak. When Derviş arrived, UNDP was still in the process of elaborating its positions on climate change.[73] Derviş spent a "considerable amount time guiding the organization" and his senior managers through intellectual discussions to develop a UNDP position.[74] He demanded of his staff rigorous analysis on how mitigation and adaptation issues would impact development trajectories and how the burden of climate change, including financing, should be shared among states.[75] During his first year in office, while internal discussions took place, he delivered no speeches on climate change.

In fact, an evaluation of the Environment and Energy Group stated that the environment was not a "core priority for the new administrator" and was critical of the climate change activities taking place.[76] The evaluation argued that "the fit between UNDP's poverty reduction and the GEF objective of mitigating global climate change has been less than convincing."[77] It argued that adaptation was a "more natural area for UNDP to engage in than mitigation, where the benefits are largely global."[78] It noted that there was a high level of dependence on the GEF and emphasis on "going after available money rather than allocating core resources to sets of activities that are consistent with the UNDP mandate."[79] The Environment and Energy Group was driven by financing opportunities from a multilateral trust and not by the executive board's priorities.

Rhetorical change

In 2006 Derviş, and his associate administrator Ad Melkert, began to speak about climate change as a development issue. In 2006 they both highlighted a new UNDP initiative—the Millennium Development Goals Carbon Facility, which offered developing states financing for carbon emission reductions. Derviş explained that it was "formulated to assist developing countries in addressing the challenge of climate change while at the same time using carbon financing opportunities to generate alternative and additional financing for reaching the Millennium Development Goals."[80] The timing of these speeches correlates with the increased global interest in climate change, signaled in part by the release of the *Stern Review* in 2006.

In 2007 climate change became one of the top global policy priorities with the release of the fourth IPCC report, the Bali UNFCCC summit and Al Gore's movie *An Inconvenient Truth*. Derviş responded to this by stating that "global warming can't be looked at as an environmental issue anymore: it is undoubtedly a threat to human development as a whole. All development strategies must therefore account for climate-related risk."[81] He made climate change a central part of his speech to the executive board in 2007, where he outlined UNDP's three-pronged approach to climate change. This involved: 1) mainstreaming climate change into UNDP's core activities; 2) creating conditions that allow markets and the private sector to "provide effective solutions to sustainable development and climate change mitigation"; and 3) increasing the capacity of developing countries to incorporate resilience into national plans.[82] The speech was important as it signaled to the executive board the importance of climate change to UNDP for the first time.

Derviş made other speeches in 2007, as climate change became the issue du jour. These were directed at a wider public audience and had two functions. Firstly, they highlighted the importance of climate change as a development issue. For instance, in 2007 he stated that "should the pace of [climate] change accelerate further, development and adaptation could well become synonymous."[83] Immediately before the Bali UNFCCC meeting Derviş published an Op-Ed on climate change and development where he stated that a "failure to act on climate change will have grave consequences for human development in some of the poorest places in the world and it will undermine efforts to tackle poverty."[84] Secondly he used public speeches to position UNDP as an agency with expertise in climate change and development. He showcased the UNDP's expertise in climate assistance to member states at council, stating that it was "one of the largest sources of technical assistance for climate change related actions in the world, with an on-going portfolio of about US$2 billion [from GEF]."[85] He continued to focus on climate change in subsequent speeches in 2008 and 2009.[86]

Policy change

Meanwhile, in 2004 the Environment and Energy Group pioneered UNDP's first adaptation policy: the Adaptation Policy Framework. The policy sought to link climate change adaptation to UNDP's mandate. It stated that "climate change adaptation remains at the forefront of any sustainable development policy agenda."[87] The framework was highly technical, directed at experts and focused on how to develop and fund adaptation programs. It was funded by the GEF as there was no core funding made available for climate change adaptation. It was significant as it was the first adaptation activity that was not a national communication for the UNFCCC.

Then a major policy shift occurred in 2007 when UNDP published the *Human Development Report, Fighting Climate Change: Human Solidarity in a Divided World.* This was UNDP's first major report linking climate change to human development.[88] It argued that climate change was a development issue as "development progress is increasingly going to be hindered by climate change. So we must see the fight against poverty and fight against the effects of climate change as interrelated efforts."[89] The report forged a strong conceptual linkage between climate change and human development stating that climate change threatens human development by eroding "human freedoms and limiting choice."[90] It emphasized that developing countries would be the worst hit by climate change, and had the lowest carbon footprints, and thus the international community should assist them in adaptation. It stated that "human development itself is the most secure foundation for adaptation to climate change."[91] The *Human Development Report*'s main contribution was to identify how climate change would impact on the poorest of the poor, and not just "polar bears," the principal perceived victims of climate change at the time.[92]

The report came as the global community was rallying around climate change—Al Gore's movie had been released and states were preparing for the Bali UNFCCC summit. Climate change was selected as a theme for the annual *Human Development Report* because of its high global salience at the time.[93] UNDP administrator Derviş and the Human Development Report Office director both felt it "was an important issue of the moment" and were aware of the need to establish a human development approach and linkage to climate change.[94] The previous *Human Development Report* had focused on water crises and water scarcity and had thus begun to outline the link between climate change and development.

The report was an agenda-setting document because of its editorial independence and its global reach. The report was launched at a side event of the Bali UNFCCC summit where Yvo de Boer (executive secretary of the UNFCCC), Ad Melkert (UNDP associate administrator), Kevin Watkins (director of the Human Development Report Office), UN secretary general Ban Ki-moon, and environment ministers from Brazil and Nigeria spoke.[95] The report was also widely publicized in each UNDP country office as UNDP country teams launched a dialogue with government officials on climate change.[96]

The *Human Development Report 2007/2008, Fighting Climate Change: Human Solidarity in a Divided World* created a "huge awareness" of the issue-linkage between human development and climate change at the global level and within UNDP.[97] The *Human Development Report* was a "very important catalytic moment" according to one UNDP staff member, as until then climate change was not well understood outside the Environment and Energy Group.[98]

However, the report was not designed to directly influence UNDP programming on climate change.[99] It did not set out prescriptive policies or recommendations for UNDP policy and it was not intended to do so.[100] Furthermore the Human Development Report Office did not make financial gains, nor was it seeking them, by selecting climate change as a topic. There were also occasional tensions between the Human Development Report Office and UNDP's country offices over the content of the report. The report argued that all states, including developing countries, should make cuts to their carbon emissions. This was controversial and not in line with the negotiating positions of developing states, such as India, at the time.

In addition to the *Human Development Report*, UNDP published a climate change strategy in 2008 which served an internal and external audience. It outlined how to integrate climate change across UNDP and justified why UNDP was the best-positioned agency to work on climate change within the UN system. The strategy built on the *Human Development Report*, stating that UNDP's overarching goals were "to align human development and climate change management efforts by promoting mitigation and adaptation activities that do not slow down but rather accelerate socio-economic progress."[101] This goal would be realized through mainstreaming climate change in UNDP's development policies as well as through the UN, national, regional, and international programs and policies. Climate change mainstreaming within UNDP would be led by the Environment and Energy Group and a "cross-practice steering group" of governance, poverty reduction, capacity development, and gender experts who would develop programming tools in each area.[102] The Climate Change Strategy and *Human Development Report* both elaborated an issue-linkage between climate change and human development and thus a rationale for UNDP's engagement with climate change.

Mandate change

In 2008 member states endorsed UNDP's new role in addressing climate change adaptation and mitigation. UNDP renegotiated with the board its multiyear strategy, to replace the previous multiyear funding framework 2004–2008. The resulting strategy document (2008) listed five key sectors that UNDP had a role and mandate to deliver policy advice and technical assistance in. These were: poverty reduction, democratic governance, crisis prevention and recovery, and environment and sustainable development. "Promoting climate change adaptation" was listed as a subset of the UNDP's environment and energy areas.[103] The strategic plan also gave UNDP a clear and official focus on adaptation: up until 2007 UNDP had no adaptation service line.[104] Although climate change had moved into UNDP's highest-level strategy document it was not highly prioritized. At the same time "funding exploded" for climate change adaptation outside of the GEF.[105] New funding opportunities for climate change emerged between 2008 and 2009 and demand from recipient countries multiplied.[106] This provided a strong incentive for UNDP to expand its climate change portfolio and the 2008–2011 period saw major changes in rhetoric, policy, structure, and operations in line with this mandate change, and I will describe these collectively here.

From 2008 there were major structural and operational changes with a marked increase in the number of staff working on climate change in the Environment and Energy Group, outside of the GEF. Bilateral donors funded UNDP to establish new programs on adaptation, deforestation, and carbon financing.[107] To manage these programs at headquarters UNDP created new teams. In 2008, for example, the Japanese government gave UNDP US$92.1 million to implement adaptation programs in 20 African states between 2008 and 2012. This was a major grant which UNDP, in partnership with UNIDO, UNICEF, and the World Food Programme (WFP), used to establish the African Adaptation Programme, UNDP's largest adaptation program. UNDP established a team to manage the program in Senegal with oversight provided by UNDP headquarters in New York.[108] One staff member explained that "there's a lot more climate change capacity in the Environment and Energy Group and less and less in the other areas."[109] The Environment and Energy Group shifted from being predominantly GEF reliant to a more even split between GEF and other projects and policy. Structural change was intertwined with operational change and enabled by new financing.

In addition, divisions outside of the Environment and Energy Group and outside of headquarters began to establish their own climate change experts. The gender unit, for instance, established a team of three people to develop policy on the links between gender and climate change and advocate for gender equality in the UNFCCC negotiations and the climate funds.[110] UNDP also established a climate change focal point system at the regional headquarters and country office level. Each regional center was assigned several "really qualified people" on climate change.[111] Staff expertise shifted from "environment and energy, to climate change, and now to climate change mitigation and adaptation separately."[112] The Environment and Energy Group sought to train "almost every single staff member from UN resident coordinators to the environmental coordinators on the UNFCCC negotiations and carbon financing."[113] The changes that had occurred in UNDP's rhetoric and policy were now manifesting in staff capacity.

In 2009 Derviş was replaced by Helen Clark, a former prime minister of New Zealand. Clark arrived the year of Copenhagen and stated from the outset that climate change should be one of UNDP's top priorities, alongside the Millennium Development Goals. In her first speech to the executive committee in April 2009 she argued that it is "critical" to bring in the "climate change challenge into the center of the way in which we think about development."[114] Clark's rhetoric built on Derviş's: she reiterated that climate change undermined development efforts and hit the poorest worst. In addition, she outlined a role for UNDP as the "UN agency with a climate and development mandate" as it had "significant expertise in the areas of climate change and sustainable development."[115] Clark had a clear view of UNDP's priorities within its mandate. UNDP was mandated to work in four areas. Two of these—promoting democratic governance, and crisis prevention and recovery—were stepping-stones to their other priorities: poverty reduction and the Millennium Development Goals; and environment and sustainable development.[116]

Clark positioned UNDP as the UN climate change and development agency. She viewed the Copenhagen summit as an opportunity to establish new climate funds[117] and lobbied governments to reduce emissions and commit new, additional resources to cover adaptation costs of developing countries.[118] She argued for a "development deal" at Copenhagen, which would benefit developing countries as well as UNDP. She stated:

> What could be achieved at Copenhagen, including through finance mechanisms being worked on, has significant implications for development. These mechanisms could become a major new and additional source of development financing, complementing, and at some point possibly even surpassing the significance of ODA [overseas development assistance]. A new development paradigm could be in the making.[119]

Clark maintained that UNDP should have a role to play in dispersing this new climate financing, agreed upon by states at the Copenhagen and Cancún UNFCCC summits.

UNDP gained member state support for focusing on climate change as a top organizational priority. At the 2009 council meeting, Clark stated that "Making the links between Millennium Development Goals achievement and sustainable development has also led me to prioritize UNDP's support to program countries on climate issues and the ongoing negotiations for a new agreement. Development and the impact of climate change and variability cannot be treated as distinct issues. They are inextricably linked."[120] UNDP also reported on its climate change adaptation spending (a total of US$11.7 million) for the first time.[121] In the 2009 annual report, Clark stated that combating climate change was one of UNDP's top mandated priorities.[122] This was a remarkable assertion for the agency to make and a significant shift from UNDP in 2000. Donors in 2010 endorsed UNDP's position and "called upon UNDP to continue playing a central role in linking climate change to development and helping developing countries to take mitigation and adaptation measures."[123] By 2010 UNDP had established a strong linkage between climate change adaptation and human development through the *Human Development Report* and other policy and rhetorical statements and gained member state support to prioritize it. The explosion of funding opportunities for climate change also enabled UNDP to expand its operations.

Significant structural change also occurred under Clark between 2009 and 2011. In 2009 UNDP outlined the need for a "surge" in staff capacity to its executive board.[124] The associate administrator, Melkert, argued that climate change was an area of "extraordinary demand" due to preparations for Copenhagen and the hoped-for future agreement on mitigation and adaptation.[125] He stated that there will be "with no doubt the need for substantial extra capacity to support in particular the least developed countries and small island states."[126]

UNDP established climate focal points in the regional bureau at headquarters and at regional and country levels. At headquarters each regional bureau established a climate change focal point. For example, the Regional Bureau for Africa had a climate change advisor reporting directly to the bureau's director.[127] UNDP also sent 26 climate change focal points to country offices in least developed countries.[128] This was part of a concerted effort to put more staff on the ground, develop climate change programs and mainstream climate change across UNDP's work. The creation of these new positions represented a significant investment of resources and locked in previous rhetoric and policy changes.

These structural changes institutionalized climate change as a central priority for UNDP. In the words of one Environment and Energy Group staff member UNDP was:

> making a huge infrastructure investment in regional and headquarters … . That's infrastructure that will not be reversed if there is policy change tomorrow … . It's really backed up by the level of activity, backed by financing at the country level, UNDP staff and experts in so many country offices and regional centers … . It's a system wide reform and system wide change in terms of handling climate change at the organization.[129]

Clark has also lobbied states to increase their climate financing at annual board meetings and international summits—from the UNFCCC in Warsaw, Poland (December 2013); to the United Nations Conference on Small Island Developing States in Apia, Samoa (September 2014). She advocated for financing for "climate-integrated development strategies" and state commitment to operationalize and adequately finance the Green Climate Fund.[130] She argued that more was needed to meet states' commitments at Copenhagen to raise US$100 billion annually by 2020, as only US$50 million had been pledged for seed funding to the Green Climate Fund. She reiterated that climate financing should be additional to current development financing. In addition, Clark noted that many developing countries were missing out on current climate assistance because they did not have the institutional capacity to navigate a "maze of more than fifty international public funds, sixty carbon markets and some 6,000 private equity funds."[131] UNDP, she explained, was "heavily involved" in building capacity of developing countries to "expand their climate finance readiness." In 2010, for instance, UNDP had produced a "framework for climate finance" discussion paper outlining the different mechanisms through which states could access finances. Clark saw UNDP's role as advocating for an increase in climate financing and enabling developing states to access this.

In summary, Clark made climate change a high priority from the outset of her administration and made further rhetorical, policy, structural, and operational changes. She made climate change one of UNDP's top two priorities, alongside the Millennium Development Goals. She claimed that UNDP was the "UN agency with a climate and development mandate."[132] She spoke regularly on climate change in the lead-up to the Copenhagen climate summit of 2009 and positioned UNDP as a conduit for the new climate assistance funds. She also won support from UNDP's executive board to include climate change as one of UNDP's top mandated priorities. As she outlined in the UNDP 2009 annual report: UNDP's mandate was to work with countries to reduce poverty, promote democratic governance, prevent and recover from crises, protect the environment and combat climate change, and try to achieve gender equality.

Operational change

In parallel to these changes, UNDP also expanded its adaptation operations considerably. In the early 2000s it had no adaptation projects but as of April 2014 UNDP had 178 projects under development, being implemented or completed.[133] In fact, in 2012 Administrator Clark highlighted that UNDP was "the largest implementer of programmes in the UN development system, with more than US$500 million in annual delivery," which translated to support for 140 countries to address climate change in 2011.[134] UNDP's adaptation projects were mainly funded through two sources: the multilateral climate funds (namely, the Special Climate Change Fund, the Least Developed Countries Fund, GEF's strategic priority on adaptation and the Adaptation Fund) and the Japanese-funded African Adaptation Programme. This section focuses on UNDP's Kenya climate change operations to investigate if and how the trends observed at headquarters were reflected at country level. It focuses on projects and initiatives which were explicitly formulated as climate change adaptation, and not those which in retrospect are seen to have elements of climate change adaptation and mitigation. It finds that UNDP Kenya's climate change adaptation operations expanded due to increased financing and country demand.

During the 1990s UNDP Kenya did not implement any climate change projects, not even through the GEF. The Kenya GEF coordinator focused predominantly on biodiversity projects and was "very successful" at gaining funding. This was because the regional advisor for GEF Kenya was a biodiversity expert. Meanwhile, GEF climate change projects were remotely supported from New York, which was not as effective. The only exception to this was in 1998, when UNDP assisted Kenya with its national communications, or reports, to the UNFCCC. UNDP Kenya mirrors the general trend at headquarters in the 1990s. There were no climate change adaptation operations, only a few GEF mitigation and enabling activities.

In 2001 UNDP began its first explicitly designed climate change mitigation project.[135] UNDP partnered with the Kenyan Association of Manufacturers to assist Kenyan businesses to embrace energy efficiency and was funded by the GEF. GEF funding enabled UNDP to work on climate change mitigation before climate change was a high priority for donors or for recipient countries.[136] Then in 2005 UNDP began its first climate change adaptation project in Kenya, Coping with Drought and Climate Change.[137] It was a regional project, also funded by the GEF, which aimed to gather, analyze, forecast, and disseminate climate information in Kenya and other sub-Saharan African states.[138] However, the project was not managed by the UNDP Kenya office. The Kenyan Drylands Center led on technical aspects and the project was managed by UNDP New York, which had expanded its climate change adaptation capacity with the hiring of new staff.[139] In sum, UNDP Kenya implemented few climate change projects in the 2000–2005 period as it was not a high priority for the Kenyan government or UNDP.[140]

Government and UNDP priorities changed in the 2006–2011 period. In 2009 the Kenyan government developed a national climate change strategy and adaptation plan. This strategy outlined national adaptation policies across the agricultural, forestry, and other relevant sectors. Kenya also established a climate change secretariat to coordinate climate change programs.[141] UNDP Kenya, meanwhile, had produced a climate change strategy that stated new office priorities such as: "building Kenya's capacity to address climate change … [and in particular] assess climate change impact and realistic response strategies and develop and implement relevant policies and regulations; and mainstreaming climate change into core government development areas … emphasizing that climate change is not only an environmental issue."[142] The Kenyan government's and UNDP's increasing prioritization of climate change went hand in hand.

In 2008 UNDP Kenya expanded its GEF climate mitigation activities. They hired a new staff member to implement a Clean Development Mechanism project in 2008 to enhance the Kenyan public and private sectors' ability to access carbon finance.[143] UNDP Kenya also established a GEF project (US$2.4 million) on standards and labeling to reduce electricity-related carbon emissions.[144] These included an Energy Market Transformation Programme which introduced new, improved cooking stoves in small enterprises and schools (US$6.6 million) and an Access to Clean Energy Services in Kenya project (US$2.31 million). These projects signified a shift in UNDP's GEF portfolio from a focus on biodiversity towards a focus also on energy efficiency, energy savings and a reduction in carbon emissions.

UNDP also sought to expand its adaptation portfolio in Kenya by partnering with the World Bank. The World Bank had developed a large arid land resource management program in 17 districts of Kenya.[145]

As a subset of this broader program, UNDP and the World Bank established the Adaptation to Climate Change in Arid and Semi-Arid Lands project. They acquired US$6.5 million from the Special Climate Change Fund to do environmental mapping, develop early warning systems and pilot community based adaptation pilot projects.[146] The World Bank managed US$5.5 million for several districts and UNDP was entrusted with the other US$1 million to focus on one district (Mwingi). Essentially, UNDP sought to "piggy-back operational and logistical support" from the World Bank's arid lands resources management project.[147]

Kenya was also one of twenty-one countries selected for the Japanese-funded African Adaptation Programme. UNDP Kenya had US$2.6 million to build capacity in government to develop climate change policy, access climate change financing, and model the impacts of climate change between December 2010 and December 2012.[148] A UNDP employee summarized the purpose of the project as:

> [The] Africa adaptation [program] works to put in place structures, which will allow the government to first of all be organized because there is an anticipated inflow of climate change related funding from the west, from donors. So it is in the best interest of the government of Kenya to be prepared … to have the right structures in place so donors also look at the delivery structures. If you've figured out your delivery structures as a country you're much better placed and you might get funding easily.[149]

Yet this project has been delayed by almost a year due to differences in budgeting procedures between the Kenyan government and UNDP.[150]

By 2014, UNDP had secured financing for several other adaptation projects. They had a GEF-Special Climate Change Fund project to strengthen national capacity to "prevent epidemic highland malaria" in Kenya.[151] They also indicated that the Japanese would fund a new Kenyan adaptation project worth US$5.5 million, as part of the next phase after the end of the initial African Adaptation Programme.[152]

In summary, UNDP expanded into adaptation at headquarters and in Kenya in the late 2000s. Until 2005 UNDP's involvement with climate change focused on mitigation and national communications for the UNFCCC and was funded solely by the GEF. This was true for UNDP headquarters and UNDP Kenya, where all projects were GEF financed during this period. There was a dramatic change between 2006 and 2014 as climate change became a top global priority and the link between climate change and development was elaborated. New climate funds proliferated including: the Special Climate Change Fund, the Least Developed Countries Fund, the Adaptation Fund, bilateral donor funds (Japan and Norway), and private sector financing. These new funds enabled UNDP to establish adaptation programs in over 21 countries. By 2014 UNDP had various climate change projects funded outside of the GEF, in particular by the Special Climate Change Fund and the Japanese government.

Conclusion

UNDP was not established with a mandate to work on climate change or the environment but has evolved considerably since its inception. In 1996 UNDP gained a mandate to work on sustainable human development, but the environment was then deprioritized in the late 1990s and early 2000s under Malloch-Brown. As a result there was little rhetoric, policy, structure, or operational engagement with climate change between 2000 and 2007 beyond UNDP's GEF portfolio.

Climate change was made a priority by UNDP administrators from 2007 onwards. There were numerous speeches by Derviş and Clark on the issue; the formulation of the *Human Development Report*, the Climate Change Strategy and the 2008 UNDP Strategy; and new climate change positions and teams were created. At the operational level UNDP acquired funding from the Japanese for the African Adaptation Programme, from the Norwegians for UN-REDD (Reducing Emissions from Deforestation and Forest Degradation), and from various other bilateral and multilateral funds for in-country programs. This was a significant shift from the previous GEF-centered climate portfolio. In 2010 member states endorsed a new mandate for UNDP, which included climate change, and in subsequent years the organization expanded its adaptation portfolio.

What drove these changes? UNDP administrators set the strategic direction of UNDP and the board tended to follow their lead. The board endorsed Malloch-Brown's shift away from the environment and Clark's prioritization of climate change. UNDP's expansion into climate change operations was also enabled by an expansion of financing opportunities from multilateral trusts and from bilateral donors. It is unlikely UNDP would have invested so many staff resources or developed almost 200 adaptation projects if this financing was not available. What is significant is that the financing flowed through a number of channels that were not tied to the executive board. Thus even if the board had not collectively agreed to prioritize climate change UNDP could nevertheless pursue climate change. In fact, UNDP follows the same trend as IOM of increased bilateralization.

UNDP had little difficulty expanding into climate adaptation, and gaining a mandate from member states for this. This was partly due to the broad nature of UNDP's mandate and the fact that UNDP had prior expertise on environmental issues through the GEF. Furthermore, the climate strategy and the publication of the *Human Development Report* on climate change in 2008 were critical. These documents substantiated for staff, and for the outside world, a link between climate change and human development. They built on an existing issue-linkage, discussed in chapter two, and demonstrated how UNDP could assist developing countries deal with the impacts of climate change.

Notes

1 Margaret P. Karns and Karen A. Mingst, *International Organizations* (Boulder, Col.: Lynn Rienner Publishers, 2004).
2 Asuncion Lera St Clair, "The Role of Ideas in the United Nations Development Programme," in *Global Institutions and Development: Framing the World?* (London: Routledge, 2004).
3 Tamar Gutner, "World Bank Environmental Reform: Revisiting Lessons from Agency Theory," *International Organization* 59 (2005): 773–783; Daniel L. Nielson and Michael J. Tierney, "Delegation to International Organizations: Agency Theory and World Bank Environmental Reform," *International Organization* 57 (2003): 241–276; Susan Park, *World Bank Group Interactions with Environmentalists: Changing International Organisation Identities* (Manchester: Manchester University Press, 2010).
4 Olav Stokke, *The UN and Development* (Bloomington, Ind.: Indiana University Press, 2009); Stephan Klingebiel, *Effectiveness and Reform of the United Nations Development Programme (UNDP)* (London: Frank Cass Publishers, 1999); Craig N. Murphy, *The United Nations Development Programme, a Better Way?* (Cambridge, Mass.: Cambridge University Press, 2006).
5 EPTA was created in 1949 by ECOSOC to provide assistance to developing countries. ECOSOC resolution 222 (IX), 1949.
6 The Special Fund was administered by ECOSOC and the General Assembly. See Karen A. Mingst and Margaret P. Karns, *The United Nations in the 21st Century* (Boulder, Colo.: Westview Press, 2007).
7 Stokke, *The UN and Development*, 187.
8 Bob Reinalda, *Routledge History of International Organizations: From 1815 to the Present Day* (New York: Routledge, 2009).
9 Ibid., 458.
10 Ibid., 458.
11 Ronald A. Manzer, "The United Nations Special Fund," *International Organization* 18, no. 4 (1964): 780.
12 General Assembly, *The Capacity of the United Nations Development System* (New York: United Nations, 1970); Reinalda, *Routledge History of International Organizations: From 1815 to the Present Day*, 459.
13 Ibid.

14 Clair, "The Role of Ideas in the United Nations Development Programme," 181–184.
15 Murphy, *The United Nations Development Programme, a Better Way?*, 232.
16 Klingebiel, *Effectiveness and Reform of the United Nations Development Programme (UNDP)*, 104.
17 Interview with former senior policy advisor at the UN System Chief Executives Board (CEB), UN Secretariat, 12 October 2010, New York. Formally UNDP is on equal standing with other UN funds and programs. However, in addition to its own programming, it supports broader UN system activities at the country level, and the administrator of UNDP chairs the UN Development Group at headquarters.
18 Morten Boas and Desmond McNeill, *Multilateral Institutions: A Critical Introduction* (London: Pluto Press, 2003), 113.
19 Murphy, *The United Nations Development Programme, a Better Way?*, 242.
20 General Assembly resolution 57/264, 2003; General Assembly resolution 49/123, 1995.
21 Desmond McNeill, "Human Development: The Power of the Idea" *Journal of Human Development* 8, no. 1 (2007), 19.
22 Ibid.
23 Mark Malloch-Brown cited in Murphy, *The United Nations Development Programme, a Better Way?*, 242.
24 Klingebiel, *Effectiveness and Reform of the United Nations Development Programme (UNDP)*.
25 Boas and McNeill, *Multilateral Institutions: A Critical Introduction*.
26 Interview with UNDP member state representative, 11 October 2010, New York.
27 Telephone interview with former UNDP Environment and Energy Group official, 4 June 2012. Note that UNDP also carried out programs in renewable energies in the 1980s. For instance, it supported China in the development of clean coal technology and helped initiate national energy conservation efforts in Peru.
28 World Commission on Environment and Development, *Our Common Future (the Brundtland Report)* (Oxford: Oxford University Press, 1987).
29 UNDP Evaluation Office, *Evaluation of the Role and Contribution of UNDP in Environment and Energy* (New York: UNDP, 2008), 157.
30 Mingst and Karns, *The United Nations in the 21st Century*, 216.
31 Zoe Young, *A New Green Order? The World Bank and the Politics of Global Environment Facility* (London: Pluto, 2002).
32 Telephone interview with former UNDP Environment and Energy Group official, 4 June 2012; Interview with UNDP Environment and Energy Group official, 22 May 2012, New York.
33 Murphy, *The United Nations Development Programme, a Better Way?*
34 Interview with UNDP Environment and Energy Group official, 6 October 2010, New York.
35 Interview with UNDP Environment and Energy Group official, 8 October 2010, New York.
36 Ibid.
37 Ibid.

38 Telephone interview with former UNDP Environment and Energy Group official, 4 June 2012.
39 Ibid.
40 Telephone Interview with former UNDP Administrator, James Gustave Speth, 14 June 2012.
41 Ibid.
42 Ibid.
43 Klingebiel, *Effectiveness and Reform of the United Nations Development Programme (UNDP)*, 106–107.
44 Note that this division went through a series of name changes in the 1990s and 2000s. It is now the Environment and Energy Group, which is how I refer to it in the next section of this chapter.
45 Telephone interview with former UNDP Environment and Energy Group official, 14 October 2010.
46 Ibid.
47 Murphy, *The United Nations Development Programme, a Better Way?*, 271.
48 Interview with UNDP Environment and Energy Group official, 20 October 2011, New York. Another claimed that GEF work made up 80 percent of their division's time while sustainable development policy work was only 20 percent. Telephone interview with former UNDP Environment and Energy Group staff member, 14 October 2010.
49 Telephone interview with former UNDP staff member, 14 October 2010.
50 Young, *A New Green Order? The World Bank and the Politics of Global Environment Facility*, 85.
51 Telephone interview with former UNDP Environment and Energy Group official, 4 June 2012
52 Interview with UNDP staff member, 8 October 2010, New York.
53 Climate change was given equal weight within UNDP's GEF portfolio alongside the other priorities. Telephone interview with former UNDP staff member, 18 October 2010.
54 Mark Malloch-Brown, *The Unfinished Global Revolution* (London: Penguin Group, 2011), 119.
55 Ibid., 121.
56 Ibid., 119.
57 Ibid., 121.
58 "Update on the UNDP Business Plans—Report to First Regular Session of the Board," UNDP (DP/2001/CRP2), 2001, 2.
59 UNDP Office of Evaluation, *Evaluation of UNDP Contribution to Environmental Management for Poverty Reduction: The Poverty-Environment Nexus* (New York: UNDP, 2010), vii.
60 UNDP Evaluation Office, *Evaluation of the Role and Contribution of UNDP in Environment and Energy* (New York: UNDP, 2008), 11.
61 UNDP, *Update on the UNDP Business Plans, 2000–2003* (New York: UNDP, 2000), 2.
62 Telephone interview with former UNDP Environment and Energy Group staff member, 18 October 2010.
63 Telephone interview with former UNDP Environment and Energy Group official, 4 June 2012.
64 Telephone interview with former UNDP Environment and Energy Group official, 4 June 2012.

65 Telephone interview with former UNDP Environment and Energy Group official, 4 June 2012.
66 Reliance on GEF funding for environment and energy initiatives meant that global environmental issues took precedence over national objectives and concerns such as pollution and water supply. The 2008 UNDP Evaluation was particularly critical of the GEF unit in UNDP for this reason. See: UNDP Evaluation Office, *Evaluation of the Role and Contribution of UNDP in Environment and Energy.*
67 UNDP Evaluation Office, *Evaluation of UNDP Partnership with Global Funds and Philanthropic Foundations* (New York: UNDP, 2012).
68 Ibid., 12.
69 Executive Board of the UNDP/Unfpa, *Report of the Executive Board on Its Work During 2004* (New York: UN, 2004).
70 Telephone interview with former UNDP Environment and Energy Group official, 4 June 2012.
71 Telephone interview with former UNDP Environment and Energy Group official, 4 June 2012.
72 Interview with UNDP official, 7 October 2010, New York.
73 Interview with UNDP official, 7 October 2010, New York.
74 Interview with UNDP official, 7 October 2010, New York.
75 Interview with UNDP official, 7 October 2010, New York.
76 UNDP Office of Evaluation, *Evaluation of UNDP Contribution to Environmental Management for Poverty Reduction: The Poverty-Environment Nexus*, ix.
77 Ibid., x.
78 Ibid., x.
79 Ibid., x.
80 Kemal Derviş 2006, *Kemal Derviş's Statement to the Executive Board of UNDP/UNFPA*, http://reliefweb.int/report/indonesia/statement-kemal-dervis-executive-board-undpunfpa.
81 Kemal Derviş 2007, *Kemal Derviş on Climate Change*, www.tr.undp.org/content/turkey/en/home/presscenter/pressreleases/2007/04/09/response-to-ipcc-report-on-climate-change.html.
82 Kemal Derviş, Speech at the Second Session of the Executive Board, New York: United Nations, 10 September 2007.
83 UNDP, "Derviş Responds to the Latest IPCC Report on Climate Change," April 3, 2007, www.tr.undp.org/content/turkey/en/home/presscenter/pressreleases/2007/04/09/response-to-ipcc-report-on-climate-change.html.
84 Kemal Derviş, *Op-Ed: Climate Change and Development: The Central Challenge of Our Time,* UNDP, New York: United Nations, 28 November 2007.
85 Derviş, *Statement by Kemal Derviş at the Second Regular Session of the UNDP/UNFPA Executive Board,* 10 September 2007, http://content.undp.org/go/newsroom/2007/september/Derviş-statement-executiveboard-20070911.en?categoryID=1005974&lang=en.
86 Derviş, "Climate Change: Reflections on the Challenge of Public Good Provision and Development," *WIDER Annual Lecture 11,* UNU-WIDER: Helsinki, 6 March 2008.
87 The Adaptation Policy Framework defined "adaptation as a process by which individuals, communities and countries seek to cope with the

consequences of climate change" and explained that while adaptation is not new, but what is "innovative is the idea of incorporating future climate risk into policy-making" UNDP, *Adaptation Policy Framework* (New York: UNDP, 2004).

88 UNDP, *Human Development Report 2007/2008* (New York: United Nations, 2008).

89 Ibid., vi.

90 Ibid., 7.

91 Ibid., 172.

92 Interview with former UNDP Human Development Report official, 12 June 2012, Oxford.

93 Interview with former director of the Human Development Report official, 3 February 2012, Oxford.

94 Interview with former director of the Human Development Report official, 3 February 2012, Oxford.

95 UNDP 2007, *Climate Change and Human Development: Charting a New Course through International Cooperation—a Panel Debate at the UNFCCC Bali*, http://hdr.undp.org/en/reports/global/hdr2007-8/launch/title,4956,en.html.

96 Interview with UNDP Energy and Environment Group senior official, 22 May 2012, New York.

97 Interview with UNDP Energy and Environment Group senior official, 22 May 2012, New York.

98 Interview with UNDP Energy and Environment Group senior official, 22 May 2012, New York.

99 Telephone interview with former Human Development Report official, 13 September.

100 Interview with former director of the Human Development Report Office, 3 February 2012, Oxford.

101 UNDP, *Climate Change Strategy* (New York: UNDP, 2008), 7.

102 Ibid., 20.

103 UNDP, *UNDP Strategic Plan 2008–2011, Accelerating Global Progress on Human Development* (New York: Executive Board of UNDP and UNFPA Annual Session, 2008), 34–35.

104 Ibid.

105 Interview with UNDP Environment and Energy Group senior official, 22 May 2012, New York.

106 Interview with UNDP Environment and Energy Group senior official, 22 May 2012, New York.

107 UNDP, FAO and UNEP initiated the UN-REDD program in 2008. REDD, or REDD+, is a financial incentive mechanism under the UNFCCC and stands for Reduction of Emissions due to Deforestation and forest Degradation in developing countries. The UN-REDD program is funded by four major donors including the Norwegian government. The UNDP staff comprises twelve technical and policy staff at the global level (based in New York, Oslo, and Geneva), two regional technical advisors for the Asia-Pacific region, and one regional advisor each in Latin America and the Caribbean and Africa. The program is worth around US$119 million. Interview with UNDP Environment and Energy Group official, 21 October 2010, New York.

108 UNDP 2015, Africa Adaptation Programme, www.undp-aap.org/. Interview with UNDP Environment and Energy Group official, 6 October 2010 and 21 May 2012, New York. Telephone interview with UNDP official, 1 June 2012.
109 Interview with UNDP official, 7 October 2010, New York.
110 Interview with UNDP gender official, 21 October 2010, New York. The gendered impacts of climate change were included in a number of speeches by the administrator and the deputy administrator. See: Ad Melkert, 2008, Climate-related conflict and the Millennium Development Goals, speech at the UNDP, *ECOSOC Luncheon on Climate-Related Conflict and the Millennium Development Goals*, http://content.undp.org/go/newsroom/2008/june/amelkert-statement-environment-conflict-mdg.en?categoryID=1486741&lang=en.
111 Interview with UNDP official, 12 October 2010, New York.
112 Interview with UNDP official, 12 October 2010, New York.
113 Interview with UNDP Environment and Energy Group senior official, 22 May 2012, New York.
114 Helen Clark, Statement on the Occasion of the Annual Session of the Executive Board of UNDP/UNFPA, 26 May 2009.
115 Helen Clark, Remarks by Helen Clark "Between Now and 2015: Moving the Development Agenda Forward," 31 August 2009, The Hague.
116 Ibid.
117 She highlighted this in conversation with the researcher at the UNFCCC summit at the UNFCCC, Copenhagen, 2009. She later stated that "Where I want more focus and action now is on … environment and sustainable development. This is particularly important … as the climate change negotiations enter an intensive period with considerable potential to benefit development." Helen Clark 2009, *Helen Clark: UN Conference on the World Financial and Economic Crisis*, http://content.undp.org/go/newsroom/2009/june/helen-clark-un-conference-on-the-world-financial-and-economic-crisis.en?categoryID=1684491&lang=en.
118 Helen Clark 2009, *Helen Clark Remarks at Pacific Leaders Event, Copenhagen*, http://content.undp.org/go/newsroom/2009/december/helen-clark-remarks-at-pacific-leaders-event-copenhagen.en?categoryID=1684491&lang=en.
119 Helen Clark, *Helen Clark: Statement at the Annual Session of the Executive Board*, 2009, http://content.undp.org/go/newsroom/2009/may/helen-clark-statement-at-the-annual-session-of-the-executive-board.en?categoryID=1684491&lang=en.
120 UNDP, *Administrators Report to the Executive Board in UNDP Reports on Sessions* (New York: UNDP, 2009).
121 UNDP, *Annual Report 2009* (New York: UNDP, 2009), 15.
122 Ibid.
123 UNDP Executive Board, *Report of the Executive Board on its work during 2010* (New York: ECOSOC), 3.
124 Ad Melkert 2009, *Speech to the UNDP/UNFPA Executive Board*, http://content.undp.org/go/newsroom/2009/june/ad-melkert-to-the-undp-unfpa-executive-board-.en?categoryID=1684491&lang=en.
125 Ibid.
126 Ibid.

127 Interview with UNDP Climate Change Focal Point in Regional Bureau for Africa, 21 October 2010.

128 Interview with UNDP Climate Change Focal Point in Regional Bureau for Africa, 21 October 2010.

129 Interview with UNDP Environment and Energy Group official, 20 October 2011, New York.

130 Helen Clark 2012, "Why Tackling Climate Change Matters for Development," www.undp.org/content/undp/en/home/presscenter/speeches/2012/11/08/helen-clark-why-tackling-climate-change-matters-for-development-/.

131 Ibid.

132 Helen Clark, Remarks by Helen Clark "Between Now and 2015: Moving the Development Agenda Forward," 31 August 2009, The Hague.

133 UNDP EEG, *Climate Change Adaptation Bulletin. A Quarterly Update of Activities—issue no. 13*, 2013, 1. UNDP EEG, *Climate Change Adaptation Bulletin. A Quarterly Update of Activities—issue no. 14*, 2013, 1. UNDP EEG, *Climate Change Adaptation Bulletin. A Quarterly Update of Activities—issue no 15*, 2014. UNDP also developed an on-line database of all their adaptation projects: the "Adaptation Learning Mechanism" UNDP 2015, *Adaptation Learning Mechanism*, www.undp-alm.org/.

134 Helen Clark 2012, "Statement of Helen Clark UNDP Administrator—Annual Meeting of the UNDP Executive Board," Geneva, www.undp.org/content/undp/en/home/presscenter/speeches/2012/06/25/helen-clark-annual-meeting-of-the-undp-executive-board/.

135 Telephone interview with UNDP official, 25 March 2011.

136 UNDP staff in hindsight reported that many of their environmental projects in the 1990s "qualified" as climate change projects. A GEF project to expand protected forestry area systems and protect biodiversity could have been labeled as a climate change project "because you're talking about advancing carbon sequestration, which starts in the forest." Interview with UNDP official, 30 March 2011, Nairobi.

137 Telephone interview with UNDP official, 25 March 2011.

138 UNDP, *Kenya: Adapting to Climate Change in Arid and Semi-Arid Lands—UNDP Project Document, Pims 3792* (Nairobi, Kenya: UNDP, 2008), 63.

139 Telephone interview with UNDP official, 25 March 2011.

140 Furthermore, UNDP and many other donors did not partner with the Kenyan government because of perceived corruption in government at the time. The election of Mwai Kibaki in 2002 and the departure of President Daniel Toroitich arap Moi put Kenya back in favor with donor governments and agencies. The GEF was unlikely to grant funds to what was widely seen as a corrupt government before then. Telephone interview with UNDP official, 25 March 2011.

141 The office has six staff members (two seconded from the Kenyan Meteorological Department, one from the department of remote sensing and remote survey, two from the environmental directorate and one from the National Environmental Authority). Interview with Kenyan climate change secretariat official, 31 March 2011, Nairobi.

142 UNDP Kenya Country Office, *Climate Change Strategy* (Nairobi, Kenya: UNDP).

143 The project was part of the UNDP-UNEP Partnership on Climate Change launched at the UNFCCC meeting in Nairobi in 2006. UNDP, *UNDP Kenya Portfolio of Energy and Environment Project/Programs 2010* (Nairobi, Kenya: UNDP, 2010), 1. See also: UNDP, *Regional CDM capacity building project for sub-Saharan Africa*, www.ke.undp.org/content/kenya/en/home/operations/projects/environment_and_energy/regional_CDM.html.

144 UNDP, *Energy and Environment Brief* (Nairobi, Kenya: UNDP, 2010). Interview with UNDP Kenya official, 30 March 2011, Nairobi.

145 The Kenya Arid Lands Resources Management, www.worldbank.org/projects/P100762/kenya-arid-lands-resource-management-project-emergency-additional-financing?lang=en.

146 UNDP, *Kenya: Adapting to Climate Change in Arid and Semi-Arid Lands—UNDP Project Document, Pims 3792*, 24–26.

147 Interview with UNDP Kenya official, 30 March 2011, Nairobi.

148 UNDP, Africa Adaptation Programme, www.ke.undp.org/content/kenya/en/home/operations/projects/environment_and_energy/Africa_Adaptation_Programme.html and interview with Kenyan climate change secretariat official, 31 March 2011, Nairobi.

149 Telephone interview with UNDP official, 25 March 2011.

150 Interview with Kenyan climate change secretariat official, 31 March 2011, Nairobi. Interview with the African Adaptation Project Manager, 14 March 2011, Nairobi.

151 UNDP, *Adaptation Learning Mechanism*, www.adaptationlearning.net/project/piloting-climate-change-adaptation-protect-human-health-kenya.

152 Ibid.

5 Conclusion

Moving beyond their mandates?

- **Comparing organizational change**
- **Did they move beyond their mandates?**
- **Implications for the study of international organizations**
- **Future research: where to from here?**
- **Policy implications**
- **Conclusion**

States created a range of international organizations, including UNHCR, UNDP, and IOM, in the aftermath of World War II. They did not intend these organizations to deal with climate change, an issue which was not on the global agenda in the 1950s but is one of the highest-priority global issues today. This book began by asking how intergovernmental institutions are adapting to climate change issues. Are they moving beyond their mandates? In this conclusion I compare the findings from the three case studies. I draw out how staff shaped the extent, timing, and nature of change in UNHCR, UNDP, and IOM. I explain how and why IOM and UNDP expanded their mandates, with member state support, but UNHCR did not. Finally, I examine the broader implications of this book for theories of international institutions and policy making on climate change, refugees, development, and migration.

Comparing organizational change

Timing of change

UNDP was the first organization covered in this book to engage with climate change. It was present at UNFCCC meetings in the 1990s and was one of three initial implementing agents of the GEF, which financed climate mitigation and later adaptation. UNDP became more involved with climate adaptation in the mid-2000s, as adaptation became a more central part of the climate regime, and it publicized its first climate adaptation policy in 2004. Meanwhile, IOM engaged with environmental issues in the 1990s through a series of conferences, reports, and research papers examining the links between the environment and migration. There was a hiatus in this work between 2000 and 2007. When climate change became one of the top global issues in the mid-2000s IOM became involved again in examining the links between climate change and migration. They published research reports, held expert conferences and lobbied states to discuss the issue. Staff in IOM and UNDP perceived a linkage between climate change and their respective mandates for migration and development.

UNHCR was initially the most reluctant to engage of the three organizations. UNHCR staff pointed out that "climate refugee" was not a legally correct term, as refugees only included those forced to flee across borders based on specific grounds of persecution. Staff expressed concern that debates over climate refugees would also distract from their obligation to ensure international protection for refugees. The High Commissioner, however, was more expansionist and linked climate change to displacement in 2007, as will be detailed next.

Extent of change

UNHCR, UNDP, and IOM all made autonomous changes to their rhetoric, policy, structure, and operations. They all sought to link climate change to their issue-area and did so without a delegated mandate from their member states. However, there were significant differences in the extent of change across these four dimensions. Let us look at each dimension of change in turn.

In terms of rhetorical change, all three had executive heads who spoke out on climate change. The High Commissioner for Refugees, António Guterres, played a prominent role in advocating for an expanded mandate in UNHCR's case. He outlined to states how climate change was a new "mega-trend" that would lead to increased displacement, and pushed states unsuccessfully to give UNHCR a new protection role. IOM's director generals also spoke about the linkage between climate change and migration, outlined IOM's expertise in this area and encouraged states to prioritize it. However, unlike Guterres, IOM's director generals did not instigate their organization's engagement with climate change. Meanwhile in UNDP, administrators differed greatly in their championing of climate change and environmental issues. Mark Malloch-Brown deprioritized the environment, but subsequently Kemal Derviş and Helen Clark reprioritized it. Clark made climate change one of UNDP's top two priorities, claiming that UNDP was the UN agency with a "climate and development mandate." In sum, executive heads perceived the importance of climate change differently and thus prioritized it to varying extents. Rhetorical change played a different role in the three organizations: in UNHCR it came first while in IOM and UNDP it followed policy change.

All three organizations issued policy statements explicitly on climate change between 2008 and 2010. UNHCR wrote a climate policy brief where they outlined how climate change could lead to further displacement. They emphasized that climate change would predominantly lead to internal displacement and those who fled across international borders due to natural disasters, with some exceptions for stateless peoples, did not fall under UNHCR's mandate. IOM in 2009 published its first policy brief explicitly on climate change and migration. Building on previous research and reports, they emphasized the "complex" relationship between climate change and migration but suggested that global migration flows would increase significantly over the next decades due to climate change. These were both internally important policy papers as they elaborated institutional positions, but it is not clear if they had any significant impact on mandate change.

UNDP's most significant policy statements were the 2007 *Human Development Report, Fighting Climate Change: Human Solidarity in a Divided World* and the 2008 *Climate Strategy*. The *Human Development Report* argued that climate change was a development issue as "development progress would increasingly be hindered by climate change."[1] It argued that human development was the most secure foundation for adaptation to climate change. Meanwhile the climate strategy outlined how UNDP would align its "human development and climate change goals" to accelerate socioeconomic progress. UNDP's contributions had the biggest impact on public debate—the *Human Development Report* has a large following and was launched all around the world. Furthermore, these policy changes created a strong rationale for UNDP to engage with climate change and led to mandate expansion.

In terms of structural changes, both UNHCR and IOM established climate change focal points, which worked closely with the IASC Task Force on Climate Change. However, UNHCR's focal point was not as well supported institutionally as in IOM—the role was passed around to three people in different departments. UNHCR and IOM had other staff also working on climate change, but neither created a significant new department especially dedicated to the issue. UNDP hired the most new staff, and invested the most resources in climate change, especially under Clark. Clark established 26 climate change focal points in country offices as well as focal points in each regional bureau in the New York headquarters. This was part of a concerted effort to develop climate change programs and mainstream climate change across UNDP's work.

UNDP arguably made the most expansive operational changes, supported by generous donor states and because they had direct access and expertise gaining grants from the GEF. Japan granted UNDP US$5.5 million for the African Adaptation Programme to build capacity in 21 African governments to develop climate change policy, access climate financing, and model the impacts of climate change. IOM also had an extensive array of over 500 climate change and environment related programs, many showcased in their 2009 compendium. These covered a broad range of interventions—from assisting Turkana pastoralists to grow crops in northern Kenya to reforestation in Haiti—and not all were clearly linked to their migration expertise. Finally, UNHCR made the fewest changes in their operations. Staff did not report assisting people who were displaced across borders due to climate change, although they did expand their natural disaster operations and assisted internally displaced peoples in affected countries.

What does this tell us about the changes that IOM, UNHCR, and UNDP made? UNDP made the most significant changes across all four dimensions and had strong reasons for doing so, as will be discussed subsequently. While IOM expanded their climate change portfolio, this did not amount to prioritizing climate change as one of their top two goals. Meanwhile UNHCR made the fewest changes.

We should also note that these four dimensions do not capture all types of change instigated by these organizations. IOM strongly lobbied states at the UNFCCC to include migration as an adaptation strategy to climate change. UNDP also lobbied donor states for greater climate financing to assist developing states. Future research could examine and compare what impact their lobbying efforts had on the climate change negotiations, and what strategies were most effective.

Nature of change

All three organizations did push for a role in climate change adaptation governance. UNDP positioned itself as the lead UN agency linking adaptation and human development. Under Clark it made upstream policy recommendations to build states' capacities to adapt to climate change and access climate financing. In addition, it implemented climate adaptation programs itself and with partner organizations like the World Bank. Thus it was both a policy advisor and operational executor.

IOM also took a role in advising governments on climate migration. They held various conferences, including the International Dialogue on Migration, and wrote research papers and reports for this purpose. Alongside this they developed their operational capacity to do adaptation projects, such as those discussed previously in Kenya. They demonstrated to states how they were a "gap filler" and could provide adaptation services when other international organizations did not. In parallel, they expanded their humanitarian operations and provided assistance in a number of natural disasters.

Meanwhile, UNHCR focused on its protection mandate. It sought to create a new international protection framework for those displaced by climate change through the Nansen Initiative (2011) and in the ministerial meeting in 2011. Although it was unsuccessful at gaining a new mandate, UNHCR continues to work with the Nansen Initiative in Geneva. UNHCR, like IOM, also expanded its natural disaster assistance. However this was typically separated out from its climate change work.

Did they move beyond their mandates?

So did we see these organizations moving beyond their mandates? All expanded their scope of activities, regardless of whether we use the 1950s, 1960s, 1990s, or 2000s as a baseline. They did so by making speeches, writing policies, hiring new staff, and making changes to their operations to engage with climate change. They also nudged states to accept that climate change was linked to their mandates, and they had a legitimate reason for expanding.

States endorsed UNDP's role in promoting climate change adaptation. At a board meeting in 2008 they agreed to a new multiyear strategy (2008–2012) that included climate change adaptation as an official service line. They subsequently agreed to Clark's prioritization of climate change as a top organizational priority, and donors called upon UNDP to continue playing a "central role in linking climate change to development and helping developing countries to take mitigation and adaptation measures."[2] This mandate change was driven by UNDP—it had clearly substantiated a connection between climate change and human development in the 2007 *Human Development Report*, and the climate strategy. This issue-linkage was strong in the eyes of most staff and states. Also many states were themselves increasingly concerned about climate change, and smaller ones such as the Maldives were actively looking for institutional levers to focus international attention and compel the major powers to act.

IOM's member states eventually endorsed its climate change induced migration work, although they initially refused to do so in 2007. At a 2009 council meeting states agreed climate change and the consequent displacement of migrants was an area of special importance. In 2010 member states noted that climate change was an important emerging issue and should be the focus of an International Dialogue for Migration summit in 2011, IOM's top policy-level engagement with states. They also endorsed the IOM Strategy, which suggested that it was in member states' "strategic interest to ensure that IOM was tasked to specifically address" climate change and other new emerging issues.[3]

Meanwhile states restricted UNHCR expansion. Although they approved and reappointed Guterres for a second term, they curtailed his efforts to establish a more routine role for UNHCR in natural disasters. And most states did not agree that UNHCR should expand its international protection role. States were reluctant to grant rights to a new group of displaced peoples that they would then be legally obliged to uphold. Nevertheless UNHCR continues to work on climate displacement in other ways: by collaborating with some supportive states through the Nansen Initiative and offering assistance to those affected by natural disasters.

Overall, these institutions developed new initiatives for climate change without official member state support. They autonomously decided climate change was an important and relevant issue which they should engage with. It would have an impact on migration, displacement and development in developing countries, which they were mandated to work on. The difference was that member states in UNDP and IOM eventually approved this expansion but in UNHCR they did not.

Explaining differences in mandate expansion

How can we explain why states supported UNDP and IOM's expansion but not UNHCR's? The boundaries of UNDP's mandate were broad and ambiguous and so could be reinterpreted easily.[4] As the UN's leading development actor it had reason to be engaged in almost anything.[5] Furthermore, the strong issue-linkage between climate change and development provided a basis for their involvement, although this was not a linear process: one administrator and some staff did not perceive climate change to be a core part of their mandate. In fact, at the end of the 1990s it looked like UNDP was retreating from its climate change work. Yet as climate change became a more important issue and the regime broadened to encompass adaptation, UNDP saw a clear rationale to engage. UNDP was able to convince states to endorse mandate expansion by outlining a strong issue-linkage between their human development mandate and climate adaptation. UNDP administrators and the Environment and Energy Group played a core role in this process.

IOM's migration mandate was not closely monitored by states, and the bilateral nature of its work—whereby states earmarked and funded projects they wanted implemented—meant the organization was able to establish a large portfolio of climate operations without a mandate from a majority of member states. IOM partnered with individual states that were concerned about climate change, such as Greece, and saw IOM as an organization that could address some effects of it. IOM also focused much of its attention on elaborating the linkage between climate change and migration through reports and research, even if they saw this link as "complex" and not causal. By establishing this issue-linkage and outlining to states the importance of climate change for migration, they were gradually able to get acceptance for their work. However, due to IOM's bilateral, project-based funding model, this mandate expansion did not constitute a major change in the organization's outputs, as states continued to fund what they wanted.

UNHCR is the most puzzling of the three cases. Why were staff initially against mandate expansion? And why didn't states endorse UNHCR's work on climate change displacement and natural disasters? In response to the first question, many staff were reluctant to add new constituencies to their existing overburdened responsibilities for refugees. UNHCR staff did not pursue expansion as they did not see a strong issue-linkage between climate change and the international protection of refugees. Debates over climate change displacement distracted from refugee needs, and confused the special protection status that refugees were guaranteed in international law. However an expansionist High Commissioner argued that climate change was a new and important "mega-trend" and driver of displacement. He built on the emerging and strengthening substantive issue-linkage between climate change and displacement, established by NGOs, academics, and other international organizations. He campaigned for member states to expand protection to those displaced across borders by climate change, and for UNHCR to play a more regularized role in assistance to IDPs of natural disasters. This was part of a broader evolution of UNHCR into a humanitarian organization that focused on a range of displaced people, not just refugees. Ultimately UNHCR, like IOM and UNDP, lobbied states to expand their mandate.

States did not support UNHCR's efforts to expand protection to those displaced across borders by the impacts of climate change. States are often unwilling to take in existing refugees, let alone to establish a precedent that another category of peoples would be allowed safe haven in to their territories. They do not want to set a legal precedent to host on their territory more migrants and displaced peoples than they already do. In fact, we see many states trying to avoid their legal obligations to refugees: from Australia's use of off-shore processing centers to European states failure to deal with a major humanitarian crisis as asylum seekers drown while trying to cross the Mediterranean. These crises illustrate how dire the situation is. States are unwilling to live up to their basic obligations under the 1951 Refugee Convention, let alone expand international law to protect other additional groups.

Even though UNHCR did not gain a mandate to expand international protection frameworks for people displaced by climate change, they continued to work on climate change displacement. They are allying with a few supportive states through the Nansen Initiative to examine how to support those most affected by climate change. In addition, they continue to provide assistance to developing countries hit by natural disasters.

What does this suggest? Firstly, that UNHCR is expanding beyond its mandated activities and has identified supportive states (such as Norway) and partner organizations (such as the IOM through the IASC) to pursue climate change and displacement-related work. Even if states refuse to change an organization's mandate international bureaucrats may find alternative ways to pursue new issue-areas. This finding stands in contrast with scholars who argue that international bureaucrats often limit their actions based on what they believe states will support.[6] It also suggests that principal agent scholars should look at how organizations lobby for change over a long time period. What principal agent scholarship views as "agency slack" may translate over the long term into mandate change as we saw with IOM and UNDP. Principal agent scholars should look at the longer-term historical evolution of mandates, and what role agents have played in shaping significant changes.

Finally, the difference in mandate change between UNDP, IOM, and UNHCR suggests a difference in the binding quality of UNHCR's mandate. Changing UNHCR's statute or expanding the 1951 Refugee Convention would have been a much more dramatic, radical step than changing IOM or UNDP's mandate. If states expanded the refugee convention they would have set a legal obligation to do much more than they are currently willing to do, as current humanitarian crises exemplify. While UNHCR's mandate places binding legal obligations on states (especially those who sign the Refugee Convention), IOM and UNDP's mandates provide strategic direction for the organizations but do not place obligations on states. Scholars have conceptualized this as a difference in "legalization"—some international organizations have "hard" legalization (high obligation, high precision, and high delegation), while others have "soft" legalization or a complete absence of legalization.[7] UNHCR, IOM, and UNDP are clearly at different ends of this spectrum as UNHCR has much "harder" legalization than IOM or UNDP. I have argued elsewhere that UNHCR and IOM represent different types of organizations.[8] UNHCR is a normative organization due to its supervisory status over a body of international law (the 1951 Refugee Convention), while IOM is a functional, project-based organization with no such normative mandate. Let's now examine what these findings mean for scholarship on international organizations.

Implications for the study of international organizations

This book has demonstrated how international organizations are adapting and moving beyond their mandates to engage with climate change. It has made three central contributions to the literature on international organizations. Firstly it has shown how three organizations have autonomously pursued a new issue-area and in two cases successfully lobbied states to expand their mandates (UNDP and IOM). Secondly, it has illustrated how international institutions may move beyond their mandates, even in the face of member state resistance. Finally, it has suggested that international bureaucrats will expand their mandate, when they perceive a strong issue-linkage between it and a new issue-area. This section turns to these three core findings and what they imply for existing scholarship on international organizations.

Firstly, this book has illustrated how staff and leaders of international organizations can influence state preferences for mandate change. They may do this through pursuing new activities, new policies, hiring new staff, and/or making speeches substantiating the linkages between their core mandate and the new issue-area. International bureaucrats can advocate for the inclusion of new priorities and issues by setting the agenda of council or executive board meetings, and reformulating strategic goals. They may also ally with sympathetic states and or other international organizations to pursue goals in a new issue-area. Over time, they may gradually influence member states' preferences and gain mandate change. We saw this in the UNDP and IOM cases. IOM officials advocated for states to discuss climate change and migration but states did not prioritize it as a topic for discussion. IOM however persisted, and eventually gained state support for their work in this area.

Here, principal agent theorists would see an example of agency slack: international institutions sought to maximize their scope and autonomy. They would explain the difference in mandate expansion by looking at how closely states monitored and patrolled UNHCR, and the qualitative difference in the legal nature of UNHCR, IOM, and UNDP's mandates.[9] As discussed above, these are important elements of understanding why up to today states have granted UNDP and IOM a mandate change, but not UNHCR.

However, principal agent scholars often imply that there is a clear line between what is within and what is outside an organization's mandate. In doing so, they do not see how organizations build a case (or substantive issue-linkage) for a new issue, and also build expertise by making policy, structural, and operational changes. Even without a mandate from states, and in the face of clear opposition, international organizations may continue to expand. The principal agent model often suggests that delegation is a once-off, one-way act, and thus misses the subtle ways that organizations influence member states' decisions over time. In particular, this research suggests international bureaucrats may shape states' views of what issues are important and which international institutions to use to address them.[10]

In addition, this book has demonstrated how staff may work outside their mandates, even when faced with member state resistance. We saw this in the UNHCR case, as states refused to grant the agency an expanded mandate at the ministerial meeting in 2011. Yet UNHCR has continued to work on the issue of climate change and displacement with sympathetic member states through the Nansen Initiative. This book suggests that scholars should focus more on the actual practices of international institutions on the ground, and how they may pursue initiatives away from member state monitoring. Principal agent accounts should not just assume that "agency slack" is a problem to be constrained and resolved. This book presented an alternative view: agency slack may indicate an organization autonomously evolving and adapting to changing global circumstances. It demonstrated that so-called "agency slack" captures an extremely wide range of activities: some activities that states may eventually support, and some that they do not. Principal agency theory should consider if and how international bureaucrats influence states' views of their organizational mandate.

Thirdly, this book adds to our understanding of when staff favor expansion. It illustrated how issue-linkages emerge and evolve (chapter two). Epistemic communities, academics, and civil society organizations elaborated strong substantive issue-linkages between climate change and migration; and climate change and development. It argued that staff will be most supportive of expanding into a new issue which they perceive links directly and causally to their mandate. As we saw in UNHCR's case, staff did not perceive a strong issue-linkage between climate change and refugees and were reluctant to expand. The High Commissioner reframed the debate as one about climate change and displacement and sought to convince states to give UNHCR a mandate on this basis. Meanwhile staff in IOM and UNDP saw how climate change would impact on their development and migration mandates. They built on and strengthened existing issue-linkages to justify their expansion into a new issue-area.

These issue-linkages did not simply spill over from the climate regime into development, migration, and refugee regimes. It was up to the staff of international organizations to internalize, reframe, and develop these issue-linkages, and use them as a basis for expansion. This contrasts with existing explanations that focus on staff's inherent and fixed desire to expand,[11] their drive to gain additional resources,[12] or their perception of member states' support for expansion.[13] It also contrasts with functional accounts, which see a passive process where cooperation in one area spills over and leads to cooperation in other areas within a given institution.[14] Scholars should consider how issue-linkages are forged, strengthened, and institutionalized.

These cases also demonstrate the power of international organizations to shape states' views of what is normatively appropriate action. Sociological institutionalist scholars would be interested in the critical role of international organizations in elaborating substantive issue-linkages to justify expansion. UNDP and IOM staff perceived that climate change could lead to, or compound, poverty and migration, and hence warranted a response. However, climate change was not creating refugees so UNHCR staff were more reluctant to engage. International bureaucrats persuaded states to expand their activities and encompass issues related to their mandated issue-area. Overall, this book contributes to our understanding of exactly how staff in international organizations can shape global governance.

Future research: where to from here?

A number of the findings in this book merit further investigation. This book examined only three international organizations and their responses to climate change. It would be interesting and important to examine how other institutions such as WHO, UNICEF, and UN Women have pursued climate change in their respective areas. Scholars could use the same framework elaborated here: focusing on the changes in rhetoric, policy, structure, operations, and overarching mandate of these institutions to see if and how staff used new issue-linkages to advocate for mandate change. Scholars interested in pursuing a larger study could track just one element of change—such as operational change—across many more organizations. This would be easier now as many institutions have more complete records of their climate operations than when this research began. This would refine and strengthen the theoretical findings elaborated here and build a broader picture of if and how our global institutions are engaging with climate change. This is extremely important given the impact climate change is likely to have on a range of different issue-areas from health to women's rights. It would also increase the generalizability of the theoretical claims presented here.

International relations scholarship would benefit from further investigating other themes of this book. Several case studies highlighted the role of the executive head in setting the priorities of the organization, lobbying states to agree to mandate expansion and developing and implementing strategic plans. The international relations literature has not significantly examined if, when, and how executive heads matter.[15] There has been a tendency to view them as epiphenomenal.[16] Scholars would do well to examine when and under what conditions leaders can change state preferences, and how effective they are at implementing strategic plans.[17]

Historical institutionalists and scholars of path dependency will be intrigued at how mandates evolve.[18] This book implies that broad mandates (UNDP's development mandate) are more flexible and ambiguous than narrower mandates (UNCHR's refugee focus). It also suggests that the legally binding definition of a refugee, enshrined in the 1951 Convention Relating to the Status of Refugees, made it more difficult for UNHCR to change. This suggests that the *scope* and the binding *nature* of mandates are key dimensions determining if, and perhaps even how, mandates will evolve. Historical institutionalists could focus more explicitly on the different types and quality of mandates, and why some organizations may evolve and change more than others.

In addition, sociological institutionalists will be interested in how organizational culture impacts on organizational expansion.[19] These three comparative cases suggest that some organizations have a stronger culture, as staff are more strongly tied to their mandate, than others. UNHCR staff for instance, stuck closely to the convention definition of a refugee, while IOM staff were not concerned if their projects did not assist migrants. Scholars could refine indicators for the strength of organizational culture, in particular the commitment staff have for their core mandate and what this means for mandate expansion.

Scholars interested in regime overlaps and institutional interplay could pay greater attention to the evolution and strength of issue-linkages. Scholars have examined why institutions may engage or "bandwagon" on other regimes and pointed to the presence of material and ideational resources.[20] They have not sufficiently examined how the strength of the linkage between one regime and another may impact on institutional behavior. We would expect, for instance, environmental organizations to be more engaged in the climate change regime than humanitarian organizations because there is stronger, more established epistemic linkage between conservation and climate change than between climate change and humanitarian activities. This will also have spin-offs for scholarship on regime complexity and fragmentation. Would we expect to see more fragmentation in regimes which have many strong issue-linkages emanating from them or do weak linkages lead to greater fragmentation?

Scholars who have concentrated on how international secretariats manage regime overlaps will be interested in these case studies of international service institutions.[21] After all, these organizations have a set of tools at their disposal which international secretariats do not: the delivery of programs and activities on the ground. This may enable them to shape the implementation of various international agreements in ways that secretariats cannot. Overlap management is not just about balancing between competing normative texts but also about how these texts are implemented in practice.

This research also noted another important trend in global governance: the increasing bilateralization of multilateral organizations. This is occurring as more donors are earmarking their contributions to international institutions.[22] In addition, many institutions are looking for alternative sources of financing from the private sector and multilateral trust funds. The result is that even if the board (or collective of member states) sets the strategic direction, they do not provide the financing and thus cannot actually ensure that organizational activities align with their goals. But does this mean more or less autonomy for international organizations? More research should examine how organizations are dealing with a competitive marketplace, and what impact this has on their mandated activities.

Policy implications

Climate change must be dealt with in many different global, national, and local institutions, not just the UNFCCC. This book has focused on how three non-environmental international institutions have adapted to climate change. It suggests that scholars and policy makers working on global environmental governance should look beyond the core climate regime. We need to consider which institutions are tasked with assisting developing countries adapt to climate change and which institutions have access to climate financing. This is particularly important given the growth of bilateral, multilateral, and private sector climate financing. In the past not all international institutions had the same capacity to access public climate financing—UNDP for instance has greater access to the GEF than IOM or UNHCR. More broadly, humanitarian work is not typically considered adaptation to climate change, despite the connections between extreme weather events and climate change. It would not matter who had access to climate financing if all international organizations were well-resourced, however this is not the case. It is critical for states to reflect on which institutions will deliver climate adaptation and what funds they will have.

This book has suggested that there are opportunities in existing development, migration, and refugee institutions to assist those affected by climate change. Development institutions such as the UNDP can ensure that developing countries are adequately prepared for climate change. They can do this by focusing on disaster risk reduction, offering assistance with climate adaptation policies, and channeling multilateral climate financing where it is needed most. Secondly, humanitarian organizations, such as UNHCR and IOM, can assist countries hit by extreme weather events and in particular provide assistance to internally displaced persons. We already see this happening in countries such as the Philippines and Vanuatu.

And what about those displaced across international borders by the effects of climate change? Scholars and many international NGOs have argued for new legal frameworks to protect those displaced by climate change. This book suggests that untangling these causes and proving that someone is displaced solely by climate change is not necessarily that important, or useful. In most cases people are forced to flee across international borders for a number of intersecting reasons: poverty, conflict, and/or persecution. Although these problems are often compounded by drought and other natural disasters (such as the case in Somalia) climate change per se is rarely the sole cause of displacement. If we are to expand international assistance and protection to a broader range of people it should be a larger group than just those affected by climate change. Otherwise we may end up inadvertently marginalizing others, such as those displaced by natural disasters not related to climate change, such as earthquakes. In summary, advocates of new international protection frameworks should look for more inclusive categories, such as "survival migrants," rather than focusing on protecting only climate migrants or those displaced by climate change.[23]

Another critical question arising from this book is when and how should international organizations adapt to new issues? Often we assume that evolution and adaptation is a good thing. It ensures that organizations maintain their relevance to global needs and are not stuck in a bygone era. However, this book has also flagged caution: organizations should not change too much, or too fast, if it means they implement activities where they have little expertise. IOM, for instance, engaged in agricultural activities in northern Kenya and there were mixed reviews of their success on this front (although this activity is not necessarily representative of all their activities). IOM operated outside its expertise and former staff and donor states were skeptical about IOM's ability to encourage pastoralists into agriculture. UNHCR's expansion in the 1990s into development activities under Ogata was questioned on similar grounds: it was seen to be going beyond its comparative advantage.

Meanwhile, the case of UNHCR suggests that there are strong reasons for organizations sticking to their mandate. For the first time in the post–World War II era we have more than 50 million refugees, asylum seekers, and internally displaced peoples worldwide.[24] If UNHCR is to ensure they gain assistance and protection necessary, it should concentrate its efforts and not be distracted by other issues. After all, if UNHCR does not protect refugee rights then who will? States, beneficiaries, and other stakeholders will all be concerned if international institutions stretch themselves across too many tasks and thus fail to deliver on their core priorities.

However, this should not be read as an excuse not to change at all. There are also arguments for UNHCR to use its expertise and operational presence to assist other displaced peoples. Organizational expansion should be managed to ensure institutions balance their core tasks and expertise with new demands.

The effectiveness of international institutions should be the central focus for future policy-orientated research. Value for money has become a core concern of member states, given the global financial crisis. Many donor states have conducted reviews of the multilateral organizations they fund "to assess the value for money provided by each organization" with the aim of "extract[ing] the greatest possible value for money from our development budget" given the "time of tough choices."[25] Some have suggested that states should also review the mandates of international institutions on a regular basis to ensure they are fit for twenty-first century purposes.[26] These institutions could have a sunset clause and those which are not delivering would be closed down.[27]

This leads to a broader question of how can and should we manage overlapping mandates and ensure our global institutions are fit for today? Overlaps may not inherently be a problem—they offer room for states to shop for the most effective institution in a given area. It could also mean that institutions can collectively work towards common goals and deliver more effective services. For example, all UN agencies will need to work towards the 17 new global sustainability goals, which were launched at the UN in September 2015. These cut across many issues from climate change to gender equality and any one goal cannot be the task of a single institution. In fact, we already see that many UN agencies are pursuing gender mainstreaming and HIV-AIDS work, as well as climate change, regardless of whether they are peacekeepers, development experts, or dedicated to health issues or women's rights. States, civil society, and international bureaucrats all need to consider how we can ensure global institutions work collaboratively and deliver global public goods most effectively.

Conclusion

This book has charted mandate change in three international institutions, UNHCR, UNDP, and IOM, between 2000 and 2015. It finds that mandate change is a gradual, iterative process and influenced by international bureaucrats. International organizations can influence states' views of what new tasks should be included in their mandate. Staff do this by building on existing issue-linkages and making changes to their rhetoric, policy, structure, and operations. As we have seen over a 15-year period, two of the three institutions in this study convinced states to give them a mandate for climate change. This book calls upon international relations scholars to examine how staff influence the trajectory and evolution of international institutions. It suggests that our institutions are adapting to deal with twenty-first-century challenges. The important question for scholars and policy-makers is how to ensure they are effective at meeting new challenges and delivering on their core mandates.

Notes

1 UNDP, *Human Development Report 2007/2008* (New York: United Nations, 2008).
2 UNDP Executive Board, *Report of the Executive Board on Its Work During 2010* (New York: ECOSOC, 2010), 3.
3 "Review of the IOM Strategy " IOM (MC/INF/302), 12 October 2010, 2.

4 Jacqueline Best, "Ambiguity and Uncertainty in International Organizations: A History of Debating IMF Conditionality," *International Studies Quarterly* 56 (2012): 674–688.
5 Craig N. Murphy, *The United Nations Development Programme, a Better Way?* (Cambridge, Mass.: Cambridge University Press, 2006), 232.
6 Andrew P. Cortell and Susan Peterson, "Dutiful Agents, Rogue Actors, or Both? Staffing, Voting Rules, and Slack in the WHO and WTO," in *Delegation and Agency in International Organizations*, ed. Darren G. Hawkins et al. (Cambridge, Mass.: Cambridge University Press, 2006), 255–280.
7 Kenneth W. Abbott, Robert O. Keohane, Andrew Moravcsik, Anne-Marie Slaughter, and Duncan Snidal, "The Concept of Legalization," *International Organization* 54, no. 3 (2000): 401–419.
8 Nina Hall, "Money or the Mandate? Why International Organizations Are Engaging with the Climate Change Regime," *Global Environmental Politics* 15, no. 2 (2015): 79–96.
9 Abbott et al., "The Concept of Legalization."
10 Compare with Joseph Jupille, Walter Mattli, and Duncan Snidal, *Institutional Choice and Global Commerce* (Cambridge, Mass.: Cambridge University Press, 2013).
11 Mark A. Pollack, *The Engines of EU Integration: Delegation, Agency and Agenda Setting in the EU* (Oxford: Oxford University Press, 2003).
12 Erica R. Gould, "Money Talks: Supplementary Financiers and International Monetary Fund Conditionality," *International Organization* 57, no. 3 (2003): 551–586.
13 Cortell and Peterson,"Dutiful Agents, Rogue Actors, or Both? Staffing, Voting Rules, and Slack in the WHO and WTO."
14 David Mitrany, *The Functional Theory of Politics* (London: London School of Economics and Political Science, 1975).
15 Michael Schroeder, "Executive Leadership in the Study of International Organization: A Framework for Analysis," *International Studies Review* 16 no. 3 (2014): 339–361.
16 Andrew Moravcsik, "A New Statecraft? Supranational Entrepreneurs and International Cooperation," *International Organization* 53, no. 2 (1999): 267–306.
17 Ngaire Woods, Nina Hall, Shubhra Saxena, Yulia Taranova, Miles Kellerman, and Hugo Batten, *Effective Leadership in International Organizations* (Geneva, Switzerland: Global Agenda Council on Institutional Governance Systems, World Economic Forum, 2015).
18 Ibid; Tine Hanrieder, "Gradual Change in International Organisations: Agency Theory and Historical Institutionalism," *Politics*, no. 34 (2014): 4.
19 Catherine Weaver, "The World's Bank and the Bank's World," *Global Governance* 13 (2007): 493–512.
20 Sikina Jinnah, "Climate Change Bandwagoning: The Impacts of Strategic Linkages on Regime Design, Maintenance, and Death," *Global Environmental Politics* 11, no. 3 (2011): 1–9.
21 Sikina Jinnah, *Post-Treaty Politics, Secretariat Influence in Global Environmental Governance* (Cambridge, Mass.: The MIT Press, 2014).
22 Erin Graham, "Money and Multilateralism: How Funding Rules Constitute IO Governance," *International Theory* 7, no. 1 (2015): 162–194.

23 Alexander Betts, *Survival Migration: Failed Governance and the Crisis of Displacement* (Ithaca, NY: Cornell University Press, 2013).
24 "World Refugee Day: Global Forced Displacement Tops 50 Million for First Time in Post-WWII Era," UNHCR, 20 June 2014. www.unhcr.org/53a155bc6.html.
25 Secretary of State for International Development Andrew Mitchell quoted in DFID, *Multilateral Aid Review: Ensuring maximum values for money for UK aid through multilateral organizations* (London: DFID, 2011), 1.
26 Oxford Martin School, *Now for the Long Term—the Report of the Oxford Martin Commission for Future Generations* (Oxford: Oxford Martin School, 2013), 58.
27 Ibid.

Select bibliography

Alexander Betts, Gil Loescher, and James Milner, *UNHCR: The Politics and Practice of Refugee Protection* (Abingdon: Routledge, 2012). An excellent overview of UNHCR's evolving mandate and operations. A must read for anyone interested in the organization.

Marianne Ducasse-Rogier, *The International Organization for Migration, 1951–2001* (Geneva, Switzerland: IOM, 2011). The only book which documents the history of the International Organization for Migration, an understudied organization. Note that the study was commissioned and published by the IOM.

Mike Hulme, *Why We Disagree about Climate Change, Understanding Controversy, Inaction and Opportunity* (Cambridge: Cambridge University Press, 2009). Hulme outlines why climate change is such a difficult problem to resolve. He argues it is an environmental, cultural, and political issue that reshapes the way we think about ourselves. He looks at it from scientific, economics, faith, psychology, communications, sociology, politics, and development perspectives.

Sikkina Jinnah, *Post-Treaty Politics, Secretariat Influence in Global Environmental Governance* (Cambridge, Mass.: The MIT Press, 2014). Jinnah demonstrates how environmnetal secretariats, such as the Convention on International Trade in Endangered Species (CITES) and the Convention on Biological Diversity (CBD), play an influential role in world politics. Secretariats manage regime overlaps, the interplay of issues, rules, and norms between international agreements. This book is part of a growing literature demonstrating the impact of global environmental organizations.

Joseph Jupille, Walter Mattli, and Duncan Snidal, *Institutional Choice and Global Commerce* (Cambridge: Cambridge University Press, 2013). This book theorizes why states "Use, Select, Change or Create" (the USCC framework) a particular international institution. It is an important contribution to international relations, as it builds bridges between historical institutionalism and rational choice theories.

Craig N. Murphy, *The United Nations Development Programme, A Better Way?* (Cambridge: Cambridge University Press, 2006). Murphy traces the evolution of the UNDP's structure, mission, ideologies, and its relationship to other institutions. It is the most detailed history of UNDP and its predecessors.

Joel E. Oestrich, ed., *International Organizations as Self-Directed Actors* (Abdingon: Routledge Global Institutions Series, 2012). A collection of chapters by leading scholars documenting the autonomy of international organizations from UNHCR to the World Bank.

Bob Reinalda, *Routledge History of International Organizations. From 1815 to the Present Day* (New York: Routledge, 2009). A useful, comprehensive overiew of the history of international organizations from the 1815 Congress of Vienna to the present day.

Index

Routledge Global Institutions Series

120 Displacement, Development, and Climate Change (2016)
International organizations moving beyond their mandates
by Nina Hall (Hertie School of Governance)

119 UN Security Council Reform (2016)
by Peter Nadin

118 International Organizations and Military Affairs (2016)
by Hylke Dijkstra (Maastricht University)

117 The International Committee of the Red Cross (2nd edition, 2016)
A neutral humanitarian actor
by David P. Forsythe (University of Nebraska–Lincoln) and Barbara Ann J. Rieffer-Flanagan (Central Washington University)

116 The Arctic Council (2016)
Governance within the Far North
by Douglas C. Nord (University of Umeå)

115 Human Development and Global Institutions (2016)
Evolution, impact, reform
by Richard Ponzio (The Hague Institute for Global Justice) and Arunabha Ghosh (Council on Energy, Environment and Water)

114 NGOs and Global Trade (2016)
Non-State Voices in EU Trade Policymaking
By Erin Hannah (University of Western Ontario)

113 Brazil as a Rising Power (2016)
Intervention norms and the contestation of global order
edited by Kai Michael Kenkel (IRI/PUC-Rio) and Philip Cunliffe (University of Kent)

112 The United Nations as a Knowledge System (2016)
by Nanette Svenson (Tulane University)

111 Summits and Regional Governance (2016)
The Americas in comparative perspective
edited by Gordon Mace (Université Laval), Jean-Philippe Thérien (Université de Montréal), Diana Tussie (Facultad Latinoamericana de Ciencias Sociales), and Olivier Dabène (Sciences Po)

110 Global Consumer Organizations (2015)
by Karsten Ronit (University of Copenhagen)

109 Expert Knowledge in Global Trade (2015)
edited by Erin Hannah (University of Western Ontario), James Scott (King's College London), and Silke Trommer (University of Helsinki)

108 World Trade Organization (2nd edition, 2015)
Law, economics, and politics
by Bernard M. Hoekman (European University Institute) and Petros C. Mavroidis (European University Institute)

107 Women and Girls Rising (2015)
Progress and resistance around the world
by Ellen Chesler (Roosevelt Institute) and Theresa McGovern (Columbia University)

106 The North Atlantic Treaty Organization (2nd edition, 2015)
by Julian Lindley-French (National Defense University)

105 The African Union (2nd edition, 2015)
by Samuel M. Makinda (Murdoch University), F. Wafula Okumu (The Borders Institute), David Mickler (University of Western Australia)

104 Governing Climate Change (2nd edition, 2015)
by Harriet Bulkeley (Durham University) and Peter Newell (University of Sussex)

103 The Organization of Islamic Cooperation (2015)
Politics, problems, and potential
by Turan Kayaoglu (University of Washington, Tacoma)

102 Contemporary Human Rights Ideas (2nd edition, 2015)
by Bertrand G. Ramcharan

101 The Politics of International Organizations (2015)
Views from insiders
edited by Patrick Weller (Griffith University) and Xu Yi-chong (Griffith University)

100 Global Poverty (2nd edition, 2015)
Global governance and poor people in the post-2015 era
by David Hulme (University of Manchester)

99 Global Corporations in Global Governance (2015)
by Christopher May (Lancaster University)

98 The United Nations Centre on Transnational Corporations (2015)
Corporate conduct and the public interest
by Khalil Hamdani and Lorraine Ruffing

97 The Challenges of Constructing Legitimacy in Peacebuilding (2015)
Afghanistan, Iraq, Sierra Leone, and East Timor
by Daisaku Higashi (University of Tokyo)

96 The European Union and Environmental Governance (2015)
by Henrik Selin (Boston University) and Stacy D. VanDeveer (University of New Hampshire)

95 Rising Powers, Global Governance, and Global Ethics (2015)
edited by Jamie Gaskarth (Plymouth University)

94 Wartime Origins and the Future United Nations (2015)
edited by Dan Plesch (SOAS, University of London) and Thomas G. Weiss (CUNY Graduate Center)

93 International Judicial Institutions (2nd edition, 2015)
The architecture of international justice at home and abroad
by Richard J. Goldstone (Retired Justice of the Constitutional Court of South Africa) and Adam M. Smith (International Lawyer, Washington, DC)

92 The NGO Challenge for International Relations Theory (2014)
edited by William E. DeMars (Wofford College) and Dennis Dijkzeul (Ruhr University Bochum)

91 21st Century Democracy Promotion in the Americas (2014)
Standing up for the Polity
by Jorge Heine (Wilfrid Laurier University) and Brigitte Weiffen (University of Konstanz)

90 BRICS and Coexistence (2014)
An alternative vision of world order
edited by Cedric de Coning (Norwegian Institute of International Affairs), Thomas Mandrup (Royal Danish Defence College), and Liselotte Odgaard (Royal Danish Defence College)

89 IBSA (2014)
The rise of the Global South?
by Oliver Stuenkel (Getulio Vargas Foundation)

88 Making Global Institutions Work (2014)
edited by Kate Brennan

87 Post-2015 UN Development (2014))
Making change happen
edited by Stephen Browne (FUNDS Project) and Thomas G. Weiss (CUNY Graduate Center)

86 Who Participates in Global Governance? (2014)
States, bureaucracies, and NGOs in the United Nations
by Molly Ruhlman (Towson University)

85 The Security Council as Global Legislator (2014)
edited by Vesselin Popovski (United Nations University) and Trudy Fraser (United Nations University)

84 UNICEF (2014)
Global governance that works
by Richard Jolly (University of Sussex)

83 The Society for Worldwide Interbank Financial Telecommunication (SWIFT) (2014)
Cooperative governance for network innovation, standards, and community
by Susan V. Scott (London School of Economics and Political Science) and Markos Zachariadis (University of Cambridge)

82 The International Politics of Human Rights (2014)
Rallying to the R2P cause?
edited by Monica Serrano (Colegio de Mexico) and Thomas G. Weiss (The CUNY Graduate Center)

81 Private Foundations and Development Partnerships (2014)
American philanthropy and global development agendas
by Michael Moran (Swinburne University of Technology)

80 Nongovernmental Development Organizations and the Poverty Reduction Agenda (2014)
The moral crusaders
by Jonathan J. Makuwira (Royal Melbourne Institute of Technology University)

79 Corporate Social Responsibility (2014)
The role of business in sustainable development
by Oliver F. Williams (University of Notre Dame)

78 Reducing Armed Violence with NGO Governance (2014)
edited by Rodney Bruce Hall (Oxford University)

77 Transformations in Trade Politics (2014)
Participatory trade politics in West Africa
by Silke Trommer (Murdoch University)

76 Rules, Politics, and the International Criminal Court (2013)
by Yvonne M. Dutton (Indiana University)

75 Global Institutions of Religion (2013)
Ancient movers, modern shakers
by Katherine Marshall (Georgetown University)

74 Crisis of Global Sustainability (2013)
by Tapio Kanninen

73 The Group of Twenty (G20) (2013)
by Andrew F. Cooper (University of Waterloo) and Ramesh Thakur (Australian National University)

72 Peacebuilding (2013)
From concept to commission
by Rob Jenkins (Hunter College, CUNY)

71 Human Rights and Humanitarian Norms, Strategic Framing, and Intervention (2013)
Lessons for the Responsibility to Protect
by Melissa Labonte (Fordham University)

70 Feminist Strategies in International Governance (2013)
edited by Gülay Caglar (Humboldt University, Berlin), Elisabeth Prügl (the Graduate Institute of International and Development Studies, Geneva), and Susanne Zwingel (the State University of New York, Potsdam)

69 The Migration Industry and the Commercialization of International Migration (2013)
edited by Thomas Gammeltoft-Hansen (Danish Institute for International Studies) and Ninna Nyberg Sørensen (Danish Institute for International Studies)

68 Integrating Africa (2013)
Decolonization's legacies, sovereignty, and the African Union
by Martin Welz (University of Konstanz)

67 Trade, Poverty, Development (2013)
Getting beyond the WTO's Doha deadlock
edited by Rorden Wilkinson (University of Manchester) and James Scott (University of Manchester)

66 The United Nations Industrial Development Organization (UNIDO) (2012)
Industrial solutions for a sustainable future
by Stephen Browne (FUNDS Project)

65 The Millennium Development Goals and Beyond (2012)
Global development after 2015
edited by Rorden Wilkinson (University of Manchester) and David Hulme (University of Manchester)

64 International Organizations as Self-Directed Actors (2012)
A framework for analysis
edited by Joel E. Oestreich (Drexel University)

63 Maritime Piracy (2012)
by Robert Haywood (One Earth Future Foundation) and Roberta Spivak (One Earth Future Foundation)

62 United Nations High Commissioner for Refugees (UNHCR) (2nd edition, 2012)
by Gil Loescher (University of Oxford), Alexander Betts (University of Oxford), and James Milner (University of Toronto)

61 International Law, International Relations, and Global Governance (2012)
by Charlotte Ku (University of Illinois)

60 Global Health Governance (2012)
by Sophie Harman (City University, London)

59 The Council of Europe (2012)
by Martyn Bond (University of London)

58 The Security Governance of Regional Organizations (2011)
edited by Emil J. Kirchner (University of Essex) and Roberto Domínguez (Suffolk University)

57 The United Nations Development Programme and System (2011)
by Stephen Browne (FUNDS Project)

56 The South Asian Association for Regional Cooperation (2011)
An emerging collaboration architecture
by Lawrence Sáez (University of London)

55 The UN Human Rights Council (2011)
by Bertrand G. Ramcharan (Geneva Graduate Institute of International and Development Studies)

54 Responsibility to Protect (2011)
Cultural perspectives in the Global South
edited by Rama Mani (University of Oxford) and Thomas G. Weiss (The CUNY Graduate Center)

53 The International Trade Centre (2011)
Promoting exports for development
by Stephen Browne (FUNDS Project) and Sam Laird (University of Nottingham)

52 The Idea of World Government (2011)
From ancient times to the twenty-first century
by James A. Yunker (Western Illinois University)

51 Humanitarianism Contested (2011)
Where angels fear to tread
by Michael Barnett (George Washington University) and Thomas G. Weiss (The CUNY Graduate Center)

50 The Organization of American States (2011)
Global governance away from the media
by Monica Herz (Catholic University, Rio de Janeiro)

49 Non-Governmental Organizations in World Politics (2011)
The construction of global governance
by Peter Willetts (City University, London)

48 The Forum on China-Africa Cooperation (FOCAC) (2011)
by Ian Taylor (University of St. Andrews)

47 Global Think Tanks (2011)
Policy networks and governance
by James G. McGann (University of Pennsylvania) with Richard Sabatini

46 United Nations Educational, Scientific and Cultural Organization (UNESCO) (2011)
Creating norms for a complex world
by J.P. Singh (Georgetown University)

45 The International Labour Organization (2011)
Coming in from the cold
by Steve Hughes (Newcastle University) and
Nigel Haworth (University of Auckland)

44 Global Poverty (2010)
How global governance is failing the poor
by David Hulme (University of Manchester)

43 Global Governance, Poverty, and Inequality (2010)
edited by Jennifer Clapp (University of Waterloo) and
Rorden Wilkinson (University of Manchester)

42 Multilateral Counter-Terrorism (2010)
The global politics of cooperation and contestation
by Peter Romaniuk (John Jay College of Criminal Justice, CUNY)

41 Governing Climate Change (2010)
by Peter Newell (University of East Anglia) and
Harriet A. Bulkeley (Durham University)

40 The UN Secretary-General and Secretariat (2nd edition, 2010)
by Leon Gordenker (Princeton University)

39 Preventive Human Rights Strategies (2010)
by Bertrand G. Ramcharan (Geneva Graduate Institute of International and Development Studies)

38 African Economic Institutions (2010)
by Kwame Akonor (Seton Hall University)

37 Global Institutions and the HIV/AIDS Epidemic (2010)
Responding to an international crisis
by Franklyn Lisk (University of Warwick)

36 Regional Security (2010)
The capacity of international organizations
by Rodrigo Tavares (United Nations University)

35 The Organisation for Economic Co-operation and Development (2009)
by Richard Woodward (University of Hull)

34 Transnational Organized Crime (2009)
by Frank Madsen (University of Cambridge)

33 The United Nations and Human Rights (2nd edition, 2009)
A guide for a new era
by Julie A. Mertus (American University)

32 The International Organization for Standardization (2009)
Global governance through voluntary consensus
by Craig N. Murphy (Wellesley College) and JoAnne Yates (Massachusetts Institute of Technology)

31 Shaping the Humanitarian World (2009)
by Peter Walker (Tufts University) and Daniel G. Maxwell (Tufts University)

30 Global Food and Agricultural Institutions (2009)
by John Shaw

29 Institutions of the Global South (2009)
by Jacqueline Anne Braveboy-Wagner (City College of New York, CUNY)

28 International Judicial Institutions (2009)
The architecture of international justice at home and abroad
by Richard J. Goldstone (Retired Justice of the Constitutional Court of South Africa) and Adam M. Smith (Harvard University)

27 The International Olympic Committee (2009)
The governance of the Olympic system
by Jean-Loup Chappelet (IDHEAP Swiss Graduate School of Public Administration) and Brenda Kübler-Mabbott

26 The World Health Organization (2009)
by Kelley Lee (London School of Hygiene and Tropical Medicine)

25 Internet Governance (2009)
The new frontier of global institutions
by John Mathiason (Syracuse University)

24 Institutions of the Asia-Pacific (2009)
ASEAN, APEC, and beyond
by Mark Beeson (University of Birmingham)

23 United Nations High Commissioner for Refugees (UNHCR) (2008)
The politics and practice of refugee protection into the twenty-first century
by Gil Loescher (University of Oxford), Alexander Betts (University of Oxford), and James Milner (University of Toronto)

22 Contemporary Human Rights Ideas (2008)
by Bertrand G. Ramcharan (Geneva Graduate Institute of International and Development Studies)

21 The World Bank (2008)
From reconstruction to development to equity
by Katherine Marshall (Georgetown University)

20 The European Union (2008)
by Clive Archer (Manchester Metropolitan University)

19 The African Union (2008)
Challenges of globalization, security, and governance
by Samuel M. Makinda (Murdoch University) and F. Wafula Okumu (McMaster University)

18 Commonwealth (2008)
Inter- and non-state contributions to global governance
by Timothy M. Shaw (Royal Roads University)

17 The World Trade Organization (2007)
Law, economics, and politics
by Bernard M. Hoekman (World Bank) and Petros C. Mavroidis (Columbia University)

16 A Crisis of Global Institutions? (2007)
Multilateralism and international security
by Edward Newman (University of Birmingham)

15 UN Conference on Trade and Development (2007)
by Ian Taylor (University of St. Andrews) and
Karen Smith (University of Stellenbosch)

14 The Organization for Security and Co-operation in Europe (2007)
by David J. Galbreath (University of Aberdeen)

13 The International Committee of the Red Cross (2007)
A neutral humanitarian actor
by David P. Forsythe (University of Nebraska) and
Barbara Ann Rieffer-Flanagan (Central Washington University)

12 The World Economic Forum (2007)
A multi-stakeholder approach to global governance
by Geoffrey Allen Pigman (Bennington College)

11 The Group of 7/8 (2007)
by Hugo Dobson (University of Sheffield)

10 The International Monetary Fund (2007)
Politics of conditional lending
by James Raymond Vreeland (Georgetown University)

9 The North Atlantic Treaty Organization (2007)
The enduring alliance
by Julian Lindley-French (Center for Applied Policy, University of Munich)

8 The World Intellectual Property Organization (2006)
Resurgence and the development agenda
by Chris May (University of the West of England)

7 The UN Security Council (2006)
Practice and promise
by Edward C. Luck (Columbia University)

6 Global Environmental Institutions (2006)
by Elizabeth R. DeSombre (Wellesley College)

5 Internal Displacement (2006)
Conceptualization and its consequences
by Thomas G. Weiss (The CUNY Graduate Center) and David A. Korn

4 The UN General Assembly (2005)
by M. J. Peterson (University of Massachusetts, Amherst)

3 United Nations Global Conferences (2005)
by Michael G. Schechter (Michigan State University)

2 The UN Secretary-General and Secretariat (2005)
by Leon Gordenker (Princeton University)

1 The United Nations and Human Rights (2005)
A guide for a new era
by Julie A. Mertus (American University)

Books currently under contract include:

The Regional Development Banks
Lending with a regional flavor
by Jonathan R. Strand (University of Nevada)

Millennium Development Goals (MDGs)
For a people-centered development agenda?
by Sakiko Fukada-Parr (The New School)

The Bank for International Settlements
The politics of global financial supervision in the age of high finance
by Kevin Ozgercin (SUNY College at Old Westbury)

International Migration
by Khalid Koser (Geneva Centre for Security Policy)

The International Monetary Fund (2nd edition)
Politics of conditional lending
by James Raymond Vreeland (Georgetown University)

The UN Global Compact
by Catia Gregoratti (Lund University)

Institutions for Women's Rights
by Charlotte Patton (York College, CUNY) and Carolyn Stephenson (University of Hawaii)

International Aid
by Paul Mosley (University of Sheffield)

Coping with Nuclear Weapons
by W. Pal Sidhu

Global Governance and China
The dragon's learning curve
edited by Scott Kennedy (Indiana University)

The Politics of Global Economic Surveillance
by Martin S. Edwards (Seton Hall University)

Mercy and Mercenaries
Humanitarian agencies and private security companies
by Peter Hoffman

Regional Organizations in the Middle East
by James Worrall (University of Leeds)

Reforming the UN Development System
The Politics of Incrementalism
by Silke Weinlich (Duisburg-Essen University)

The International Criminal Court
The Politics and practice of prosecuting atrocity crimes
by Martin Mennecke (University of Copenhagen)

BRICS
by João Pontes Nogueira (Catholic University, Rio de Janeiro) and Monica Herz (Catholic University, Rio de Janeiro)

The European Union (2nd edition)
Clive Archer (Manchester Metropolitan University)

Protecting the Internally Displaced
Rhetoric and reality
Phil Orchard (University of Queensland)

For further information regarding the series, please contact:

Nicola Parkin, Editor, Politics & International Studies
Taylor & Francis
2 Park Square, Milton Park, Abingdon
Oxford OX14 4RN, UK
Nicola.parkin@tandf.co.uk
www.routledge.com